Electrode Dynamics

A. C. Fisher

University of Bath

Series sponsor: **ZENECA**

ZENECA is a major international company active in four main areas of business: Pharmaceuticals, Agrochemicals and Seeds, Specialty Chemicals, and Biological Products.

ZENECA's skill and innovative ideas in organic chemistry and bioscience create products and services which improve the world's health, nutrition, environment, and quality of life.

ZENECA is committed to the support of education in chemistry.

OXFORD NEW YORK TOKYO
OXFORD UNIVERSITY PRESS
1996

Oxford University Press, Walton Street, Oxford OX2 6DP

Oxford   New York
Athens   Auckland   Bangkok   Bombay
Calcutta   Cape Town   Dar es Salaam   Delhi
Florence   Hong Kong   Istanbul   Karachi
Kuala Lumpur   Madras   Madrid   Melbourne
Mexico City   Nairobi   Paris   Singapore
Taipei   Tokyo   Toronto

and associated companies in
Berlin   Ibadan

Oxford is a trade mark of Oxford University Press

Published in the United States
by Oxford University Press Inc., New York

A catalogue record for this book is available from the British Library

Library of Congress Cataloging in Publication Data
(Data available)

ISBN   0 19 855690 X

Typeset by AMA Graphics
Printed in Great Britain by The Bath Press, Somerset

Series Editor's Foreword

Oxford Chemistry Primers are designed to provide clear and concise introductions to a wide range of topics that may be encountered by chemistry students as they progress from the freshman stage through to graduation. The Physical Chemistry series contains books easily recognized as relating to established fundamental core material that all chemists need to know, as well as books reflecting new directions and research trends in the subject, thereby anticipating (and perhaps encouraging) the evolution of modern undergraduate courses.

In this Physical Chemistry Primer, Adrian Fisher has produced a clear and concise introduction to the important topic of *Electrode Dynamics* assuming no prior knowledge of the area and with a minimum of mathematics. This Primer will be of interest to all students of chemistry (and their mentors).

Richard G. Compton
Physical and Theoretical Chemistry Laboratory
University of Oxford

Preface

Electron transfer plays a pivotal role in determining the outcome of chemical reactions. Electrochemistry offers a unique and powerful approach to examine these events providing key insights into the fundamental factors which drive such processes. This text sets out to provide the reader with a general overview of background theory and experimental techniques employed by electrochemists in the study of electrode reactions. Early chapters concentrate on quantifying the influence of electron transfer kinetics and mass transport via diffusion, convection, and migration, on electrode reactions. These themes are developed in Chapter 3 where the principles of voltammetry are explored and examples from the recent academic literature used to illustrate the versatility of this technique. Chapter 4 focuses on the tiny region close to the electrode surface known as the electrical double layer. In particular, models are described and experimental techniques outlined which probe this important region. The final chapter covers the important area of spectroelectrochemistry which is a field of intense academic research. As the name suggests these are techniques which combine electrochemical and spectroscopic measurements to examine the often highly complex electrode reactions. Examples are taken from the recent literature and include the application of scanning tunnelling microscopy to the investigation of nanoscale surface reactions as well as techniques such as esr spectroscopy which probe homogeneous phase reactivity.

I am indebted to many friends and colleagues who have made valuable contributions to this text. I would like to particularly thank Richard Compton, Melanie Kershaw, and Laurie Peter for their advice during the production of this text and Michael Weaver for original photographs of his STM studies of sulfur described in Chapter 5.

Bath   A.C.F.
April 1996

Contents

1   Introduction

Dynamic electrochemistry is the study of electron transfer reactions between electrodes (typically, but not always, metallic) and reactant molecules, usually in a solution phase. The subject is concerned with the development and application of techniques that probe the rate and mechanism of these reactions in much the same way that the familiar methodology of homogeneous kinetics examines chemical reactions in which atoms or molecules fragment or rearrange.

Many parameters have been found to influence the dynamics of an electrode reaction. In principle, the following can all control the rate of the charge transfer process:
- the electrode potential;
- transport of material between the electrode and bulk solution;
- the reactivity (ease of oxidation or reduction) of the solution species;
- the nature of the electrode surface;
- the structure of the interfacial region over which the electron transfer occurs.

Before examining these factors in detail important concepts from equilibrium electrochemistry are summarized.

1.1   Equilibrium Electrochemistry Revised

An electrode process often involves the transfer of charge across the interface between a metallic electrode (m) and a solution phase (aq) species, as illustrated by the following example:

$$Fe^{3+}(aq) + e^-(m) \rightleftharpoons Fe^{2+}(aq) \tag{1.1}$$

It is obvious that the above reaction can only proceed once a suitable electrode has been inserted into the solution phase which acts as a source or sink (depending on the relative concentrations of Fe^{2+} and Fe^{3+}) of electrons. Notice that, since the reaction involves the transfer of a charged particle (the electron) between the two distinct phases, (m) and (aq), as this electron transfer reaction moves towards equilbrium a net charge separation must develop between the electrode and the solution. This charge separation creates a potential difference at the electrode/solution interface (Fig. 1.1). If ϕ_s is the solution potential and ϕ_m the metal potential, then this potential

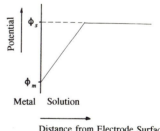

Fig. 1.1 Schematic diagram of the potential drop across the metal/solution interface

drop across the electrode/solution interface $\Delta\phi_{m/s}$ is given by

$$\Delta\phi_{m/s} = \phi_m - \phi_s$$

An implication of the development of an interfacial potential difference is that the electrochemical reaction may be sensitive to this potential gradient. As a consequence the equilibrium position of a particular reaction may be influenced by altering the potential gradient at the interface.

Although this quantity has a fixed and precise value for any particular system, a little thought indicates that direct measurement of the quantity $\Delta\phi_{m/s}$ at a single interface is not experimentally feasible. Measurement of a potential difference requires a complete, conducting, electrical circuit, and for the case in question this necessitates the introduction of at least one additional electrode. The latter, inserted in the solution, will also possess a potential drop ($\Delta\phi_{m/s}$), so that any voltage-measuring device connected to the system will monitor the difference of the potential drops at the two electrode/solution interfaces.

To address this problem electrochemists introduce a reference electrode. This is a device that maintains a constant potential drop $\Delta\phi_{m/s}$ across its interface with the solution. In this way when measurements of potential difference, E, are made between the two electrodes, the observed voltage is given by

$$E = (\phi_m - \phi_s) + \text{constant} \tag{1.2}$$

where the term in parenthesis refers to the electrode of interest and the constant describes the role of the reference electrode.

Experimentally, the investigation of equilibrium electrode processes is performed using an electrochemical cell such as that idealized in Fig. 1.2. The apparatus is composed of two half-cells, each containing a solvent (e.g. water). The species of interest (such as Fe^{3+}/Fe^{2+}) are contained in one half-cell along with a chemically inert, metal electrode (e.g. gold). The reference electrode is held in the other half-cell and the two half-cells are separated by a porous membrane (or salt bridge), forming the complete electrochemical cell. Provided no current is drawn through the cell then E rapidly reaches a steady value (E_e) corresponding to equilibrium, which under these conditions is dependent upon the relative concentrations of Fe^{3+} and Fe^{2+}. For the example shown in Fig. 1.2 the reference electrode is a standard hydrogen electrode (SHE), so that

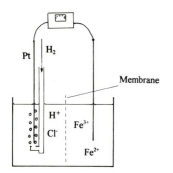

Fig. 1.2 Schematic of a simple electrochemical cell

$$E_e = \Delta\phi_{m/s}(Fe^{3+}/Fe^{2+}) - \Delta\phi_{m/s}(SHE) = \Delta\phi_{m/s}(Fe^{3+}/Fe^{2+}) - 0$$

By convention the SHE is defined as possessing an absolute potential of zero. This enables us to conveniently report potentials of other half-cells (for example, the Fe^{2+}/Fe^{3+} system) relative to this reference electrode. In particular, for the general charge transfer reaction

$$O(aq) + ne^-(m) \rightleftharpoons R(aq)$$

in which n electrons are transferred, Nernst showed that the potential established at the electrode under equilibrium conditions is given by

$$E_e = E^{\ominus} + \frac{RT}{nF} \ln \frac{[O]}{[R]} \tag{1.3}$$

where the equilibrium potential (E_e) of the electrode results from the standard electrode potential ($E^{\ominus}$) of the reaction and the concentrations of O and R at the electrode surface, which, under equilibrium conditions, are the same as their values in bulk solution.

Equilibrium electrochemical measurements enable thermodynamic parameters such as reaction free energies, entropies and enthalpies, equilibrium constants including solubility products, activities, and solution pHs to be readily obtained compared with alternative techniques. It is this area that is most often emphasized when students first encounter electrochemistry.

Strictly the concentrations in the Nernst equation should be replaced by activities. For the case shown it has been assumed that the activities of O and R are unity.

The standard electrode potential $E^{\ominus}$ is that for the case where the half cell of interest has unit activity of all its potential determining ions, and the potential is measured relative to the SHE.

1.2 Electrolysis

The electrochemical cell shown in Fig. 1.2 can also be operated under conditions where a current flows. This is achieved by the application of a potential to the cell which is different in value to E_e. The current induces the exchange of electrons between the electrode and molecules in solution, so altering the oxidation state of the molecule, and 'electrolysis' occurs. The transfer of electrons can be in either direction; a molecule in solution (e.g. Fe^{3+}) may accept an electron from the electrode and become reduced

$$Fe^{3+}(aq) + e^-(m) \longrightarrow Fe^{2+}(aq) \tag{1.4}$$

Alternatively, an electron can be removed from the molecule (Fe^{2+}) by the electrode, the molecule being oxidized

$$Fe^{2+}(aq) \longrightarrow Fe^{3+}(aq) + e^-(m) \tag{1.5}$$

In an oxidative process electrons flow from solution-phase molecules to the electrode, and vice versa for a reduction.

For the reductive electrolysis of Fe^{3+} to Fe^{2+} shown in Fig. 1.3 the magnitude of the current (i) is given by

$$i = AFj \tag{1.6}$$

where
- F is the Faraday constant (96 485 C mol^{-1}),
- A is the electrode area (cm^2),
- j is the 'flux' of Fe^{3+} reaching the electrode surface (moles cm^{-2} s^{-1}).

This last quantity, the flux j, is important and can be envisaged as the quantity of material reacting in the electrode area per second. It can be thought of as the rate of the electrochemical reaction. As such it is described by a rate law, not dissimilar in form from those commonly encountered in homogeneous kinetics

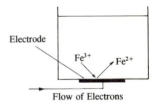

Electrode

Fe^{3+} Fe^{2+}

Flow of Electrons

Fig. 1.3 The conversion of Fe^{3+} to Fe^{2+} at a metallic electrode connected to an external circuit

When a current is allowed to pass, the measurement of electrochemical processes becomes slightly more complicated. As will be noted later, this is overcome by the use of a three-electrode system in which the current is allowed to flow between the electrode of interest and an auxiliary electrode. The potential of the electrode is then held relative to a stable reference electrode.

$$j = k_0[\text{Fe}^{3+}]_0 \qquad (1.7)$$

where k_0 is the heterogeneous rate constant for the electron transfer reaction and $[\text{Fe}^{3+}]_0$ the concentration of the reactant at the electrode surface. Notice that it has been assumed that the reaction rate is first order. This is frequently (but not always) observed.

A very significant difference arises in comparison with the equilibrium situation discussed in Section 1.1. The act of passing a current through the cell converts Fe^{3+} to Fe^{2+} at the electrode/solution interface, with the result that the concentration of Fe^{3+} at the electrode surface is no longer the same as that in bulk. This is because, typically, the rate of depletion through electrolysis is faster than the speed at which reactant may be replenished by diffusion from bulk solution to the interfacial region. For this reason the concentration term shown in eqn 1.7 explicitly relates to the surface concentration of the electroactive species.

It follows that the observed electrolytic current may be dependent on either

The relative importance of mass transport and electron transfer rates will depend on their comparative magnitudes.

- the transport of reactants to, or products from, the electrode surface (rate-determining 'mass transport').

or

- the rate (k_0) of the heterogeneous electron transfer (rate-determining 'electrode kinetics').

which are in turn controlled by the electrode potential, as discussed in the next section.

1.3   Dynamic electrochemistry

The previous section concluded that the electrode kinetics for any electrolytic reaction can be controlled by the potential. Inspection of eqn 1.7 shows that in order to study this relationship, the potential dependence of the heterogeneous rate constant for the electron transfer reaction must be quantified.

The rate of electron transfer

The aim of this section is to develop a general theory that describes quantitatively the rate of electron transfer accompanying the simple, general reaction

$$\text{O}(aq) + \text{e}^-(m) \underset{k_{\text{ox}}}{\overset{k_{\text{red}}}{\rightleftharpoons}} \text{R}(aq) \qquad (1.8)$$

in which the two chemical species are interconverted by a single electron transfer reaction at the electrode. It is assumed that there are arbitrary quantities of O and R present in solution and that k_{red} and k_{ox} describe the first-order heterogeneous rate constants for the forward (reductive) and back (oxidative) electron transfer reactions.

Using eqn 1.7 the current for the oxidative and reductive components of reaction 1.8 can be predicted

$$i_c = FAk_{red}[O]_o \qquad (1.9)$$

$$i_a = FAk_{ox}[R]_o \qquad (1.10)$$

The subscripts 'c' and 'a' for the reductive and oxidative currents, respectively, arise from the idea that a reductive current is passed at a cathode (c) and an oxidative current at the anode (a).

where $k_{red}[O]_o$ and $k_{ox}[R]_o$ are the respective fluxes of material to the electrode surface. The net current flowing (i) for the reaction is

$$i = i_a + i_c \qquad (1.11)$$

therefore

$$i = FA(k_{ox}[R]_o - k_{red}[O]_o) \qquad (1.12)$$

Note that when, and only when, reaction 1.8 is at equilibrium, so that eqn 1.3 holds, current flow has ceased and the fluxes $k_{ox}[R]_o$ and $k_{red}[O]_o$ must be balanced

$$k_{red}[O]_o = k_{ox}[R]_o \qquad (1.13)$$

See Cox (1994) for a detailed description of transition state theory.

It is reasonable to assume that electron transfer reactions behave in an analogous manner to chemical rate processes, and might therefore be described using the transition state model Fig. 1.4. This model views the reaction as occurring via a path that involves the reactants $[O_{(aq)} + e^-_{(m)}]$ overcoming an energy barrier, the summit of which is termed the transition state, *en route* to becoming products $R_{(aq)}$. Transition state theory predicts the rate of the reduction reaction (k_{red}) to be

$$k_{red} = A\exp\left(\frac{-\Delta G^\ddagger_{red}}{RT}\right) \qquad (1.14)$$

where $\Delta G^\ddagger_{red}$ is the free energy of activation and A is a 'frequency factor', which (in this application) accounts for the rate of collision of the electroactive molecule with the electrode surface.

In the example above we need to identify the activation free energy for both the forward (reductive) and back (oxidative) reactions. For a fixed temperature and pressure, the former is simply the free energy change between the reactant and transistion state $(\Delta G^\ddagger_{red})$

$$\Delta G^\ddagger_{red} = G^\ddagger - G_{ox}$$

whilst the latter is

$$\Delta G^\ddagger_{ox} = G^\ddagger - G_{red}$$

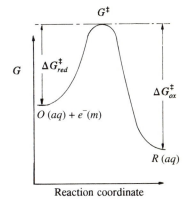

Fig. 1.4 Free energy plot for a simple one electron reduction of a species O(aq)

For the case where the reaction is in equilibrium and the bulk concentrations of O and R are equal, the corresponding plot shown in Fig. 1.4 becomes symmetrical with

$$\Delta G_{ox}^{\ddagger} = \Delta G_{red}^{\ddagger}$$

The electrical energy is charge (q) multiplied by potential (E).

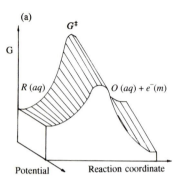

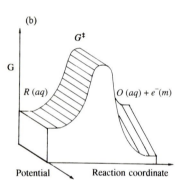

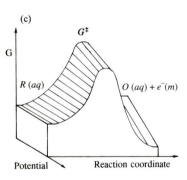

Fig. 1.5 Free energy – potential surfaces for (a) $\alpha = 0$, (b) $\alpha = 1$ and (c) $\alpha = 0.5$.

The effect of potential

The kinetic model developed thus far is entirely analogous to that familiar from homogeneous chemistry. However, from the discussion in Section 1.1 it is clear that electrochemical reactions are influenced by the interfacial potential ($\Delta\phi_{m/s}$). To establish how this alters the development of transition state theory as applied to electrode kinetics, consider the free energies for the electrode reaction

$$Fe^{3+}(aq) + e^-(m) \xrightleftharpoons{} Fe^{2+}(aq)$$

The free energy of the reactants (Fe^{3+} and e^-) is

$$\begin{aligned} G_{Fe^{3+}} &= \text{constant} + 3F\phi_s - F\phi_m \\ &= \text{constant} + 2F\phi_s - F(\phi_m - \phi_s) \end{aligned} \tag{1.15}$$

and of the product (Fe^{2+})

$$G_{Fe^{2+}} = \text{constant}' + 2F\phi_s \tag{1.16}$$

The transition state free energy may be reasonably inferred to be intermediate between eqns 1.15 and 1.16 and so can be written as

$$G^{\ddagger} = \text{constant}'' + 2F\phi_s - (1-\alpha)F(\phi_m - \phi_s) \tag{1.17}$$

The value of $\alpha(0<\alpha<1)$ reflects the sensitivity of the transition state to the drop in electrical potential between the metal and the solution. If α is close to zero then the transition state resembles the reactants in its potential dependence whereas when it approaches unity the transition state behaves in a product-like manner. In fact, α (the so-called transfer coefficient) is typically found to be close to $1/2$ for many reactions, suggesting that the transition state has intermediate behaviour.

It is evident from eqns 1.15 to 1.16 that changes in the solution potential ϕ_s, or the metal potential ϕ_m, will not only alter the *absolute* values of the free energies of the reactants, products, and transition state Fig. 1.5, but also their *relative* values. The consequence of this is that the free energy barriers to reduction (forward reaction) and oxidation (back reaction) are modified and this can be seen to be the origin of the potential dependence of the kinetics of the electrode reaction.

The above provides a basis for understanding the sensitivity of electrochemical rate constants (k_{red} and k_{ox}) to changes in electrode potential by considering the potentials at a single electrode/solution interface. In practice, as noted above, experiments are necessarily made on a cell containing more than one electrode. It will be seen in Chapter 2 that whereas two electrodes suffice for tiny electrodes (so-called microelectrodes, which pass very small currents), for larger currents and electrodes it becomes necessary to work with a three-electrode system. In either situation the potential, E, is applied between a 'working' electrode (at which the electrode process of interest takes place) and a reference electrode (cf. Fig. 1.2). Again, as in Section 1.1, the potential drop at the reference electrode/solution interface is fixed. Consequently, alterations in the applied potential, E, directly reflect changes in the interfacial potential drop at the working electrode as suggested by eqn 1.2.

It has been established that when the applied potential (E) is equal to the equilibrium potential, (E_e), no current will flow through the cell. For any other value of E, the electrode reaction 1.8 will no longer be at equilibrium with the electrode potential and so electrolysis becomes thermodynamically viable. Whether or not any current flows will depend on the *kinetics* of the particular electrode reaction under investigation.

The foregoing has demonstrated that for electrolysis to occur a potential different in value to that of E_e for the reaction in question must be applied to the working electrode in order to 'drive' the electrode reaction. The deviation of E from the equilibrium potential is given the term 'overpotential'

The term α is sometimes referred to as the symmetry factor. This arises from the idea that α effects the shape of free energy–potential plots such as those shown in Fig. 1.5.

$$\eta = E - E_e \qquad (1.18)$$

So far, the description of interfacial rates has been restricted to the consideration of a process (specifically reaction 1.8) occurring at one electrode/electrolyte interface. The relevant (eqn 1.14) and the free energy of activation specified in it were seen to depend on the difference of reactant and transition state free energies quantified by eqns 1.15 and 1.16. The form of these equations relates to a *single* interface. As such they are not particularly useful for experimental purposes, where it is much more desirable to know how k_{red} (and k_{ox}) depend on the overvoltage applied to a cell of two (or three) electrodes. The following relationships result

By defining the overpotential relative to E_e for the particular reaction studied, the resulting value has a directly comparable meaning with other equilibria and therefore comparisons of overpotentials for different systems can be more easily interpreted.

$$k_{red} = A \exp\left(\frac{\Delta G^{\ddagger}_{red}}{RT}\right) \exp\left(\frac{-\alpha F \eta}{RT}\right) \qquad (1.19)$$

$$k_{ox} = A \exp\left(\frac{\Delta G^{\ddagger}_{ox}}{RT}\right) \exp\left(\frac{(1-\alpha)F \eta}{RT}\right) \qquad (1.20)$$

where

$$\Delta G^{\ddagger}_{ox} = \Delta G^{\ddagger}_{o} - (1-\alpha)F\left(E_e - E^{\ominus}\right) \qquad (1.21)$$

$$\Delta G^{\ddagger}_{red} = \Delta G^{\ddagger}_{o} + \alpha F\left(E_e - E^{\ominus}\right) \qquad (1.22)$$

These relationships may be simplified by the introduction of the new constants k^{o}_{red} and k^{o}_{ox}, which are independent of η

$$k_{red} = k^{o}_{red} \exp\left(\frac{-\alpha F \eta}{RT}\right) \qquad (1.23)$$

The term

$$\Delta G^{\ddagger}_{o}$$

corresponds to the free energy change accompanying the reaction when $E_e = E^{\ominus}$.

$$k_{\mathrm{ox}} = k_{\mathrm{ox}}^{o} \exp\left(\frac{(1-\alpha)F\eta}{RT}\right) \tag{1.24}$$

More strictly, eqn 1.25 results not solely from combining the four earlier equations but also benefits from the simplification that results when no current flows (i=0. This provides a further relationship between k°_{red} and k°_{ox}.

The standard exchange current i_0 where

$$i_{\mathrm{o}} = FAk^{\circ}[\mathrm{R}]_{\mathrm{bulk}}^{\alpha}[\mathrm{O}]_{\mathrm{bulk}}^{1-\alpha}$$

can essentially be considered as a scaling factor which is dependent upon the experimental reactant concentrations and the value of the standard rate constant.

The Butler–Volmer equation

All that remains is to substitute eqns 1.23 and 1.24 into eqn 1.12 to give the final relationship for the net current, i, flowing at the working electrode

$$i = i_{\mathrm{o}}\left(\frac{[\mathrm{R}]_{\mathrm{o}}}{[\mathrm{R}]_{\mathrm{bulk}}}\exp\left\{\frac{(1-\alpha)F\eta}{RT}\right\} - \frac{[\mathrm{O}]_{\mathrm{o}}}{[\mathrm{O}]_{\mathrm{bulk}}}\exp\left\{\frac{-\alpha F\eta}{RT}\right\}\right) \tag{1.25}$$

This is the important Butler–Volmer equation, which is fundamental to electrode kinetics, (see Butler (1924) and Volmer (1930)). It predicts how the observed current varies as a function of the overpotential and transfer coefficient, α. If the solution under investigation is well stirred, the surface concentrations of the reactants will be equal to their bulk values i.e. ([R]$_{\mathrm{o}}$ = [R]$_{\mathrm{bulk}}$ and [O]$_{\mathrm{o}}$ = [O]$_{\mathrm{bulk}}$). Under these conditions eqn 1.25 simplifies to

$$i = i_{\mathrm{o}}\left(\exp\left\{\frac{(1-\alpha)F\eta}{RT}\right\} - \exp\left\{\frac{-\alpha F\eta}{RT}\right\}\right) \tag{1.26}$$

Next the form of the Butler–Volmer equation is explored. The variation of i is shown in Fig. 1.6, where two limiting cases are considered, corresponding to large or small values of i_{o}. In both cases, as expected, no current flows when $\eta = 0$. The difference between them arises in the current response to the imposition of small overpotentials.

When i_{o} is large 'reversible' little or no applied overpotential is required to drive the reaction and current flows readily in both anodic and cathodic directions according to the sign of the overpotential applied. Moreover, because i_{o} is large the net current, i, (at any overpotential) will have significant contributions from both i_{a} and i_{c}, except at very large positive (negative) overpotentials where the cathodic (anodic) component ultimately becomes negligible.

For processes that have a small value of i_{o} 'irreversible' a high overpotential is required to induce current flow. Fig. 1.6 shows that if the overpotential is increased to a value at which, say, the oxidative (anodic) process is driven, then the corresponding reductive component is vanishingly small. Equally, at overpotentials that drive the reductive (cathodic) process, negligible oxidation currents flow. In the former case the Butler–Volmer equation simplifies to

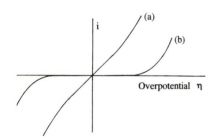

Fig. 1.6 The variation of current as a function of the overpotential for (a) a reversible and (b) an irreversible electrode reaction.

$$\ln i = \ln i_{o} + \frac{(1-\alpha)F}{RT}\eta \tag{1.27}$$

and in the latter to

$$\ln(-i) = \ln i_o - \frac{\alpha F}{RT}\eta \qquad (1.28)$$

Eqns 1.27 and 1.28 show how values α of can be found experimentally from a knowledge of the current/voltage characteristics of an irreversible electrode reaction. This is known as Tafel analysis (involves plotting $\ln(i)$ against E) and is illustrated in Fig. 1.7. Notice that if both the cathodic and anodic processes can be measured and are analysed via Tafel plots then the quantity i_o (known as the standard exchange current) may be estimated from the point of intersection of the two extrapolated Tafel lines.

Tafel (1905) analysed electrochemical data using $\ln(i)$ against E rather than η since E_e is often unknown.

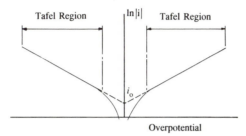

Fig. 1.7 Tafel analysis

Given that the term 'irreversible' has been linked with the idea that the electrode process, when driven, proceeds solely in one direction with no back reaction, the term 'reversible' may be readily appreciated to signify the fact that, as already noted, both the anodic and cathodic components contribute significantly at all but the most extreme overpotentials. As to whether or not reversible or irreversible behaviour is found for any particular system has been seen to depend on whether i_o is 'small' or 'large'. These are relative terms and signify the time-scale of the electrode kinetics relative to transport of material in and out of the interfacial region. This will be better appreciated after reading Chapter 2.

1.4 Overpotential and energy levels

It is useful to illustrate the above model of electrode reactions with a simple physical picture in which the effect of potential on a metallic electrode is considered.

Metals contain a lattice of closely packed atoms. The overlap of atomic orbitals within these lattices ensures that electrons in the solid are able to move freely. The electrons occupy an effective continuum of energy states in the metal, up to an energy maximum termed the Fermi Level. An electrical potential applied to a metal acts to increase or decrease the energy of the Fermi Level (Fig. 1.8).

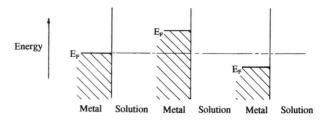

Fig. 1.8 The influence of potential on the fermi level in a metal

In Fig. 1.9 a reactant molecule, O, capable of undergoing a charge transfer reaction has been introduced. Some of the energy states of O are depicted and, in particular, it has a lowest unoccupied molecular orbital ('LUMO') at an energy higher than that of the metal's Fermi Level. It is evident that for electrons to leave the metal to occupy the LUMO (and so electroreduce O) is, in this case, an unfavourable process. However, if the energy in the metal is raised by the application of a negative electrical potential, such that the electrons in the metal have a higher energy than in the LUMO, then electron transfer becomes thermodynamically favourable. Furthermore if the electrode potential is made more and more negative eqn 1.15 predicts that the free energy of activation for electron transfer from the electrode to the metal will become smaller and the rate of reduction of O will increase. Thus, it can be seen that the increase in k_{red} as the overpotential becomes negative correlates with the shift in the energy of the Fermi level in the electrode relative to that of the LUMO of O.

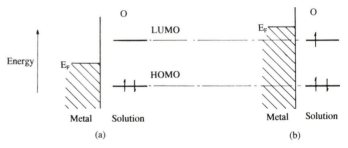

Fig. 1.9 Electron transfer and energy levels (a) the electrode potential is insufficient to drive the reduction of O and (b) at a more reductive potential where the electrode process becomes thermodynamically favourable.

1.5   Marcus theory of electron transfer reactions

In the last section two extreme classes of electrode process were described — reversible and irreversible — according to whether fast or slow electrode kinetics operate. It is of interest to enquire as to the origin of these different patterns of behaviour at the molecular level. A widely adopted model has been reported by Marcus (1963).

Consider the electroreduction of O to R, as discussed above. This electron transfer, which occurs via quantum mechanical tunnelling of electrons from the electrode to O, is thought to take place on a time-scale of 10^{-15}–10^{-16}s. In contrast, nuclear motions (vibrations) within O (or its solvation shell) occur on the significantly longer time-scale of 10^{-13}s. It follows that when electron transfer occurs the product R must still possess the same molecular shape (bond lengths, bond angles, etc.) and solvation shell structure as pertained to the reactant O the instant before the electron transferred. The electron transfer is thus constrained to follow the Franck–Condon Principle.

A second constraint also operates. As no loss or gain of energy accompanies the electron transfer, R must be formed with an energy that exactly matches the sum of the energy of the electron in its Fermi Level in the metal and the energy of O, both measured the instant preceding electron transfer.

The simultaneous operation of both constraints implies that, for electron transfer to take place, the O molecule must become energetically excited and the R molecule formed is also energetically excited. This activated state of O is the transition state (see p. 5) for the reaction, since it represents the structure in which O and R have the same geometry (shape and solvation) and satisfy the energy conservation restriction identified above.

The transition state for the electroreduction under consideration is seen to involve the energetic activation of O into a geometry that is intermediate between the equilibrium geometries of (unexcited) O and R.

The energy required for this activation is likely to be greater the more gross the difference in molecular geometry (bond angles, bond lengths, degree of solvation, orientation and position of local solvent dipole moments, etc.). Accordingly, the (free) energy of activation is smaller (and the rate of electroreduction as measured by k_{red} larger), the less the change in molecular geometry accompanying the reduction, where the term 'geometry' is understood to embrace both bond angle and bond length changes ('inner sphere' effects) and changes in solvation.

A simple rule applicable to simple electron transfer reactions ('outer sphere', free from adsorption effects) follows: fast electrode processes will be observed when both species, O and R, in the redox process have comparable shapes and solvation. For example the reduction of MnO_4^- to MnO_4^{2-} is reversible, as both R and O are tetrahedral in shape and have closely similar bond lengths. Moreover, since the charge is extensively delocalized their solvation is not dissimilar.

The transfer of electrons between electrode and O involves the quantum mechanical tunnelling of electrons between the two locations. This is highly sensitive to the distance between the two locations and necessitates that O is no more than about 10 Å from the electrode surface.

Bibliography

Bard A. J. and Faulkner L. R. (1980), *Electrochemical Methods, Fundamentals and Applications*, Wiley, New York.

Bockris J. O'M. and Reddy A. N. (1970), *Modern Electrochemistry*, Plenum, New York.

Bond A. M. (1980), *Modern Polarographic Methods in Analytical Chemistry*, Dekker, New York.

Butler J. A. V. (1924), *Trans. Faraday Soc.* **19**, 734.

Cox B. G. (1994), *Modern Liquid Phase Kinetics*, Oxford University Press.

Crow D. R. (1988), *Principles and Applications of Electrochemistry*, Chapman and Hall, London.

Erdey Gruz T. and Volmer M. (1930), *Z. Physik. Chem.*, **150A**, 203.

Marcus R. A. (1963), *J. Phys. Chem.* **67**, 853.

Rieger P. H. (1987), *Electrochemistry*, Prentice-Hall International, Englewood Cliffs, NJ.

Tafel J. (1905), *Z. Physik. Chem.* **50A**, 641.

Southampton Electrochemistry Group (1985), *Instrumental Methods in Electrochemistry*, Ellis-Horwood, Chichester.

2 Mass transport

Chapter 1 considered the rate of an electron transfer reaction occuring at the electrode/solution interface. However, it is clear that other physical processes may contribute to the overall kinetics of any particular reaction. In particular, it can be appreciated that in order for electrolysis to proceed the reactant molecule must be transported from the bulk solution to the electrode interface, and that the reaction products, once formed, will themselves diffuse away from the electrode. A basic summary of the steps involved is shown in Fig. 2.1.

The overall reaction rate will be limited by the slowest step. A particular reaction might therefore be controlled either by the kinetics of electron transfer or by the rate at which material is brought to (or from) the electrode through mass transport (e.g. diffusion).

It was revealed in the previous chapter that the electron transfer rate can be experimentally controlled through the electrode potential and that it typically varies by several orders of magnitude for a relatively small increment in the potential. It may therefore be anticipated that when electrolytic reactions are induced to occur rapidly, mass transport effects may become important for the successful interpretation of experimental observations.

According to experimental conditions the following transport processes are found to be significant

- diffusion
- convection
- migration

The significance of each of the transport properties noted above will now be considered in a little more detail.

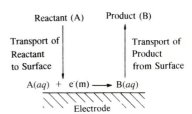

Fig. 2.1 Schematic of some of the processes that influence the rate of an electrode reaction in which A is reduced to B.

2.1 Diffusion

Diffusion arises from uneven concentration distributions and acts to maximize entropy by smoothing out inhomogeneities of composition within any system.

The rate of diffusion at a given point in solution is dependent upon the concentration gradient at that particular location. Fick (1855) first described diffusion mathematically by considering the simple case of linear diffusion to a planar surface (Fig. 2.2). He showed experimentally that the number of moles of material diffusing through a unit area in one second, the 'diffusional flux', j, for a species B is

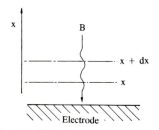

Fig. 2.2 Diffusion to a large planar surface.

The quantity called the flux was first discussed in Chapter 1.

$$j = -D_B \frac{\partial [B]}{\partial x} \tag{2.1}$$

where [B] is the concentration of B. The constant of proportionality, D_B, is known as a diffusion coefficient and is characteristic of the diffusing species. Equation 2.1 is Fick's first law of diffusion.

In practice, electrochemists are more usually concerned with the change in concentration at a point (such as that adjacent to the electrode surface) as a function of time. This can be established by considering the variation in concentration of material within the region x to $x + dx$ (Fig. 2.3) during a time interval (dt). This is simply related to the difference in flux of B entering through the plane at x and the flux leaving through the plane at $x + dx$ during that time. This can be expressed by

$$[B]_{(x,t+dt)} A dx - [B]_{(x,t)} A dx = j_{(x,t)} A dt - j_{(x+dx,t)} A dt \tag{2.2}$$

which is a statement of mass conservation.

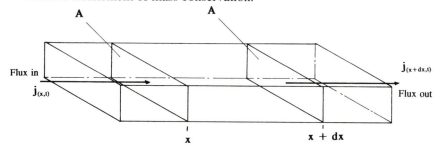

Fig. 2.3 Schematic of the flux of a species into and out of a zone bounded by two planes separated by a distance *dx*.

Simple rearrangement of eqn 2.2 results in the general conservation equation

$$\frac{\partial}{\partial t}[B]_{(x,t)} + \frac{\partial}{\partial x} j_{(x,t)} = 0 \tag{2.3}$$

Substitution of the diffusional flux from eqn 2.1 into eqn 2.3, gives the desired relationship

Equations of the form 2.4 are readily solved using numerical computational approaches. For further details see Britz (1988).

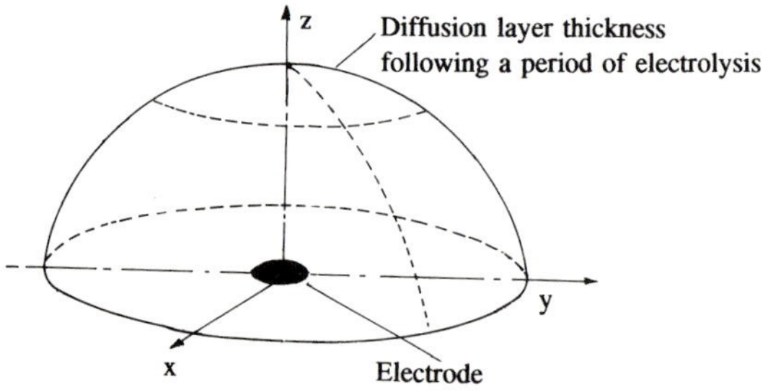

$$\frac{\partial [B]}{\partial t} = D_B \left(\frac{\partial^2 [B]}{\partial x^2} \right) \tag{2.4}$$

z

Diffusion layer thickness following a period of electrolysis

y

x

Electrode

Fig. 2.4 The diffusion field around a microdisc electrode after a short period of electrolysis

This differential equation is Fick's second law of diffusion. It enables the prediction of concentration changes of electroactive material close to an electrode surface. When more than one direction needs to be considered, for example in three Cartesian directions (x, y, z), the diffusional equation must be extended to describe transport in each direction (Fig. 2.4)

$$\frac{\partial[\text{B}]}{\partial t} = D_{\text{B}} \frac{\partial^2[\text{B}]}{\partial x^2} + D_{\text{B}} \frac{\partial^2[\text{B}]}{\partial y^2} + D_{\text{B}} \frac{\partial^2[\text{B}]}{\partial z^2}$$

This is derived in a similar manner to eqn 2.4.

2.2 Convection

Movement due to convection occurs when a mechanical force acts on a solution. Two types of convection may be usefully distinguished.

The first is natural convection, which can be present in any solution and arises from thermal gradients and/or density differences within the solution. In electrolysis reactions, the former may arise simply as a result of the exo- or endo-thermicity of the process, and the latter as a result of the electrode reaction creating concentrations of products near the electrode of different density to those of the reactants in bulk solution. Natural convection typically becomes a significant perturbation in electrolysis conducted with electrodes of conventional size (approximately mm or larger) on the time-scale of 10–20 s and longer, and is generally undesirable since it is difficult to predict.

The second type of movement is forced convection, which is achieved by external forces such as gas bubbling through solution, pumping, or stirring.

In some electrochemical experiments an element of forced convection is sometimes deliberately introduced. This has the function of swamping any contributions from natural convection, so ensuring that reproducible experiments can be made over time-scales beyond the 10–20 s limit. In particular, the forced convection is usually arranged to possess well-defined hydrodynamic behaviour. This enables a quantitative description of the flow in solution to be established and the pattern of mass transport to the electrode to be rigorously predicted. Mathematically, concentration changes resulting from movement of solution with a velocity v_x are given by

$$\frac{\partial[\text{B}]}{\partial t} = -v_x \frac{\partial[\text{B}]}{\partial x} \tag{2.5}$$

This equation is the convection analogue of eqn (2.4).

2.3 Migration

The external electric field $(d\phi/dx)$, which exists at the electrode/solution interface as a result of the drop in electrical potential between the two phases $(\Delta\phi_{m/s})$, is capable of exerting an electrostatic force on charged species present in the interfacial region and thereby of inducing the movement of ions to or from the electrode. These 'migration' effects contribute to transport phenomena accompanying electrochemical reactions.

The migratory flux (j_m) is proportional to the concentration of the ion, [B], the electric field, and the ionic mobility (u)

The electrostatic force apparent on an ion under an external electric field ϕ is given by

$$\text{Force} = \frac{zF}{N_A}\frac{\partial \phi}{\partial x}$$

$$j_m \propto -u[B]\frac{\partial \phi}{\partial x} \tag{2.6}$$

The ionic mobility is dependent on the ionic charge and size, as well as on the solution viscosity.

2.4   Experimental elimination of migration effects

The interplay of migration and electrolysis in any electrochemical reaction gives rise to a complex physical transport process which is hard to control or interpret. For example, consider a charge transfer process occurring at an electrode surface. The consequence of this will be a change in the concentration of ionic species local to the electrode and the result is a change in the electrical potential in the solution. Hence, the electric field ($d\phi/dx$), near the electrode is also altered. This, in turn, causes a change in migratory fluxes (eqn 2.6) and the rate of mass transport of electroactive material, both to and from the electrode, will change as the electrolysis proceeds. This varying rate of mass transport makes interpretation of experimental data difficult, to say the least.

To simplify this problem experiments are often performed under conditions where migration can be neglected. This is made possible by means of a chemically and electrochemically inert 'background electrolyte' added to the solution in high concentrations: levels around, or in excess of 0.1M may be commonly used.

The way in which the background electrolyte operates can be understood by examining electrolytic ion injection or removal at the electrode/solution interface. This causes a slight redistribution of the different cations and anions comprising the background electrolyte to take place near the interface, so as to maintain near electrical neutrality in all of the interfacial region – except for a very tiny region absolutely adjacent to the electrode (see below). The availability of background ions to maintain electroneutrality ensures that electric fields, ($d\phi/dx$), do not build up in the solution as electrolysis proceeds. In this way transport effects are almost completely blind to the small charge injection that arises from electrolytic reaction. This means that the migratory mass transport effects detailed in eqn 2.6 can be neglected.

For further details regarding the effects of mass transport in electrode systems the reader is directed to Brett and Brett (1986).

Calculations have shown that approximately 100 times the concentration of background electrolyte relative to that of the reactant is required if the measured current is not to be perturbed significantly (<1%) from its value where migration plays no part.

The addition of background electrolyte also provides several other highly important benefits. First, a high concentration of electrolyte increases the solution conductivity making it less resistive to the flow of current; otherwise, the passage of current through the cell may be simply limited by the bulk solution conductivity as opposed to the nature of the interfacial reaction of interest. Secondly, the large excess of background electrolyte controls the distance at the interface over which the electrical potential drops from its value in the metal electrode (ϕ_m) to that in bulk solution (ϕ_s). This so-called 'double layer' (see Chapter 4) is compressed, by the high levels of electrolyte used into a very small region of approximate dimensions 10–20Å, over which the full interfacial drop in potential ($\Delta\phi_{m/s}$) occurs. This has two vital consequences: (i) the absence of potential gradients outside this narrow region prevents migration effects as discussed above, and (ii) the distance is short enough to be compatible with tunnelling of electrons between the metal electrode and the reactant in solution. The latter is essential if a reagent is to get sufficiently close to the electrode *and* experience the full driving force ($\Delta\phi_{m/s}$) for the electron transfer reaction.

A third advantage is that the ionic strength of the solution stays effectively constant during electrolysis because of the relatively high concentration of background electrolyte compared to that of the reactants and products. This means that the activity coefficients of the reactants and products (which are highly sensitive to ionic strength) are kept constant. As a result (i) the electrode potentials necessary to bring about electrolysis, as predicted from the Nernst equation, and (ii) the rate of any chemical reaction of the electrode products (or reactants), are both constant throughout the experiment.

Experimentally the so-called 'formal potential' is recorded when experiments are performed under conditions of constant solution activity.

The activity of a species x (a_x) is defined as
$$a_x = \gamma_x [X]$$
where γ_x is the activity coefficient.

2.5   Transport in electrolysis

Transport properties are considered next, in relation to the simple electrochemical oxidation of B to B$^+$:

$$B(aq) \longrightarrow B^+(aq) + e^-(m) \tag{2.7}$$

In the presence of excess background electrolyte, the above process can be viewed as comprising the movement of B through solution up to a point within 10–20 Å of the electrode surface, where the drop in potential between the electrode and the solution induces the transfer of the electron (by tunnelling) from B into the metal electrode. Outside this region, it has been noted that negligible gradients of electric potential exist, so that the material must be transported to this site of electron exchange by diffusion and/or convection alone, with migration playing no part.

Of particular interest are the concentration distributions of B and B+ in space (concentration 'profiles') adjacent to the electrode surface (outside the 10–20 Å double layer) developed when reaction 2.7 occurs. Two cases are considered: in the first the reactant B is transported to the electrode by diffusion and in the second transport occurs via a combination of *diffusion* and *forced convection*.

Diffusional transport

Consider a planar electrode placed in a large container of a reactant B, Fig. 2.5. Before the electrode is connected to a potential source the solution

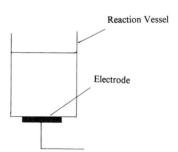

Fig. 2.5 Schematic representation of a large planar electrode embedded in one wall of a vessel containing a reactant B

composition is constant and a uniform bulk concentration of B pertains throughout the solution. If the electrode is then connected to a potential source, which drives reaction 2.7 to completion, any B at the electrode interface is converted to B^+. As a result, a concentration gradient is induced between the electrode (where the concentration of B approaches zero) and bulk solution (where its value is unchanged from the initial value). The concentration gradient so created is perpendicular to the electrode (x-direction) and acts to force a flux of fresh B from bulk solution to flow towards the electrode surface (as described by eqn 2.1). Gradually a 'diffusion layer' is established close to the electrode in which the concentration of B differs from its value in bulk solution. As electrolysis proceeds the thickness of this layer becomes progressively larger. Figure 2.6 shows the concentration profile at certain times after electrolysis has been initiated. It can be seen that close to the electrode surface the profile is linear, but beyond this it approaches the bulk concentration value asymptotically.

The scale of the diffusion layer grows steadily as the electrolysis consumes more and more B. In principle, the diffusion layer thickness can grow without limit until it has exhausted all the B within the container, and this is the case if the transport effects experienced by B are purely diffusional in nature. The resulting spreading out of the diffusion layer with time is illustrated in Fig. 2.6.

> By assuming a very large electrode we have been able to consider material flux in the x-direction only and to ignore contributions from other directions.

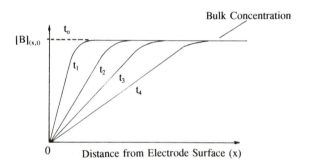

Fig. 2.6 The growth of the diffusion layer thickness as a function of time (t).

In reality the bulk of the solution invariably experiences some mixing, maintained by natural convection, and this limits the scale to which the diffusion layer can expand. Obviously, the greater the natural convection in bulk, the thinner the final diffusion layer thickness and the sooner after commencement of electrolysis is a constant diffusion layer thickness established. In other words, the concentration profile evolves to a steady-state diffusion layer thickness controlled by the extent of the mixing Fig. 2.7 in bulk solution. Under these steady-state conditions transport *within* the diffusion layer is pictured as occurring via diffusion alone, whilst outside the layer the concentration of B is maintained at its bulk value through natural mixing.

Nernst estimated the size of the diffusion layer thickness (δ) by extrapolation of the linear region in Fig. 2.8, close to the electrode, until it reaches a

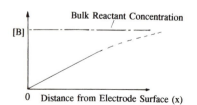

Fig. 2.7 Concentration profile once steady state has been established.

concentration corresponding to that in bulk solution. Under normal experimental conditions, for electrodes with millimetre dimensions, the Nernst diffusion layer thickness is approximately 0.05 cm.

Nernst's approach also enables an estimate of the flux of material to the electrode surface from which the current, due to linear diffusion perpendicular to the electrode, is readily found

$$\frac{i}{FA} = j_d = D_B \frac{[B]_{bulk} - [B]_o}{\delta_d} \qquad (2.8)$$

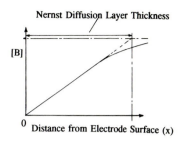

Fig. 2.8 Nernst diffusion layer thickness.

$[B]_o$ is the concentration of B at the electrode surface. This is controlled by the electrode potential; in Figs. 2.6 and 2.7 a strongly oxidizing electrode potential has been assumed so that the surface concentration of B is indistinguishably close to zero.

Mass transport by convection and diffusion

In systems where there is convection in addition to diffusion extra material is brought to the electrode surface and the currents that flow are larger than those observed when diffusion alone operates. Nevertheless, it is conventional to retain the notion of a diffusion layer and the use of eqn 2.8. The 'diffusion' layer thickness, δ_d, is smaller than before and now depends on the rate of convection. Needless to say, transport to the electrode is by both diffusion and convection, but the effect of the latter is allowed for in the convection dependence of δ_d, as will become clear later in this chapter.

Transport-limited current

Since electroactive material must be transported to the electrode surface for electrolysis to occur, it follows that the maximum observable current is limited by the rate at which reactant reaches the electrode/solution interface. From eqn 2.8 it is clear that if the electrode potential is gradually increased so that the surface concentration of the reactant ($[B]_o$) eventually becomes zero, a limiting current i_L is reached

$$i_L = \frac{D_B FA[B]_{bulk}}{\delta_d} \qquad (2.9)$$

Under these conditions all the electroactive material transported to the electrode is converted to products.

2.6 Hydrodynamic electrodes

A hydrodynamic electrode is one in which forced convection is deliberately introduced to dominate transport to the electrode. Convection within the electrochemical cell may be induced in several ways. The electrode may be fixed and electrolyte solution allowed to flow across the surface, or the electrode may move (for example by rotation) inducing convection in the solution.

For simplification of the associated theoretical description, experiments are usually designed so that laminar flow is obtained. In laminar flow the

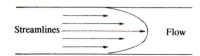

Fig. 2.9 Laminar flow along a thin tube.

The hydrodynamic properties may be characterized by a number of different parameters including the Schmidt and Peclet numbers. For further reading in this area the reader is directed to the article by Brett and Brett (1986) or Oldham (1986).

solution moves smoothly in layers (laminae) along constant directions (streamlines) as shown in Fig. 2.9. This contrasts with the irregular, chaotic motion that characterizes turbulent flow. Prediction of whether flow within an electrochemical cell is turbulent or laminar in nature is readily accomplished using the Reynolds number (*Re*) concept. This is a dimensionless parameter which is related to the cell geometry and the solution velocity through

$$Re = \frac{(\text{length}) \times (\text{velocity})}{(\text{kinematic viscosity})}$$

where the length and velocity characterize the electrode size and the convective flow as illustrated later. For any particular hydrodynamic electrode there exists a critical value of *Re* above which the flow changes from laminar to turbulent.

Hydrodynamic electrodes permit experimental variation of the transport characteristics, for example, by changing the solution flow rate. Controlling the rate of convection fixes the time spent near the electrode surface by species brought to, or formed at, the surface. This is important when electron transfer forms chemical reaction intermediates of specific lifetimes. It is possible to adjust the convection so that the time period in which the electrode 'sees' these intermediates is either long or short compared to the time-scale of their existence. Comparison of the electrode responses in these two limits then permits the lifetime of the intermediates to be deduced. Convective control in this way fixes the *time-scale* of the electrochemical experiment. This important concept is encountered in detail in Chapter 3.

The rotating disc electrode

A commonly encountered hydrodynamic electrode is the rotating disc electrode (RDE) which is particularly popular because it is simple to use and easy to construct. A small, metallic disc (often Au or Pt, of radius about 0.5 cm) acts as the working electrode and is embedded centrally within a large cylinder composed of an insulating material such as Teflon. The cylinder is placed into a vessel that contains the solution of interest (Fig. 2.10) and rotated at a constant speed. The rotation spins the solution out from the cylinder surface in a radial direction and this movement in turn, draws fresh material up towards the disc. This sustains a steady supply of electroactive material to the electrode.

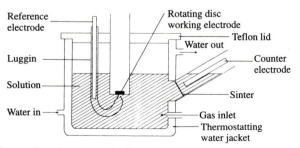

Fig. 2.10 The rotating disc electrode, RDE

The Reynolds number for the RDE is given by

$$Re = \frac{\omega r_c^2}{v}$$

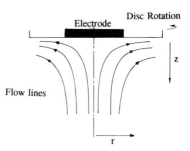

Fig. 2.11 Laminar flow lines at the rotating disc.

where r_c is the radius of the cylinder, ω the rotation speed in Hz (cycles per second), and v the viscosity of solution. Provided Re does not exceed about 1×10^5 the flow induced in the solution is laminar and has the pattern shown in Fig. 2.11. The convective flow sketched can be quantified precisely. Of particular importance is the component of the flow normal to the electrode, which is responsible for bringing fresh material to the surface. The velocity component (v_z) in this direction, close to the electrode surface ($z \longrightarrow 0$) has been shown to be

$$v_z \approx -Cz^2$$

where $C = 0.510\omega^{3/2}v^{1/2}D^{-1/3}$ and where η is the solution viscosity. Notice that as the rotation speed is increased the velocity close to the surface (v_Z) also increases, as expected.

To predict the flux of electroactive material, B, to the electrode surface both diffusion and convection must be considered. Under steady-state mass transport conditions

$$\frac{\partial[B]}{\partial t} = D_B \frac{\partial^2[B]}{\partial z^2} - v_z \frac{\partial[B]}{\partial z} = 0 \qquad (2.11)$$

It can be appreciated that this equation is a composite of eqns 2.4 and 2.5, which separately describe diffusion and convection. Note that, in accord with standard experimental practice, sufficient background electrolyte is assumed to be present to permit the simplification outlined in Section 2.4 to neglect effects due to migration. Equation 2.11 can be solved assuming [B] is zero at the electrode surface, corresponding to the transport-limited current, and the [B] found as a function of z. The limiting current is then readily evaluated from $\partial[B]/\partial z$ through the use of eqn 2.9. The result is

$$I_L = 0.62nFA[B]_{bulk} D_B^{2/3} {}_Bv^{-1/6}\omega^{1/2} \qquad (2.12)$$

This is the Levich equation for the RDE and predicts the variation in the transport limited current (i_L) as a function of the rotation speed. In particular, a plot of the transport-limited current against the square root of rotation speed should yield a straight line which passes through the origin (Fig. 2.12).

It is informative to consider why the limiting current increases with rotation speed. As noted above, the velocity of the solution towards the electrode is greater at higher rotation speeds. This leads to enhanced convective transport of the electroactive material and a concomitant shrinking of the diffusion layer thickness as predicted in Section 2.5. Quantitatively,

$$\delta_d = 1.61D_B^{1/3}v^{1/6}\omega^{-1/2}$$

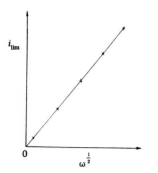

Fig. 2.12 A Levich plot for an RDE experiment.

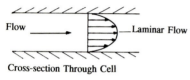

Cross-section Through Cell

Fig. 2.14 Parabolic flow profile in the channel electrode.

where ν is the kinematic viscosity of the solution (viscosity/density). The influence of the diffusion layer thickness on the transport-limited current is, following the discussion in Section 2.5 (p.17), given by

$$j_L = \frac{D_B[B]_{bulk}}{\delta_d} \qquad (2.13)$$

The behaviour shown in Fig. 2.12 is now readily rationalized, since the diffusion layer thickness is seen to vary inversely as the square root of the rotation speed. As δ_d shrinks, the concentration gradient of B, which drives a flux of B to the electrode, gets steeper and enhanced currents are predicted by eqn 2.13.

The channel electrode

As was noted earlier, forced convection can also arise from the movement of electrolyte solution over a stationary working electrode. One example of such a system is the channel electrode (Fig. 2.13). This employs a thin rectangular duct through which solution is mechanically pumped. The electrode is embedded smoothly in one wall of the duct. If laminar flow is established the velocity profile across the channel is parabolic with the maximum velocity occurring at the channel centre (Fig. 2.14).

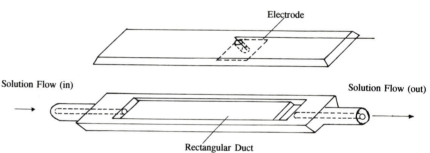

Fig. 2.13 The channel electrode.

The convective-diffusion equation for mass transport within the rectangular duct under steady state conditions is given by

$$D_B \frac{\partial^2[B]}{\partial y^2} - v_x \frac{\partial[B]}{\partial x} = 0 \qquad (2.14)$$

Solution of this equation enables the prediction of the mass transport limited current as a function of the solution volume flow rate (V_f) in an analogous fashion to the RDE:

$$i_L = 0.925nF[B]_{bulk} D_B^{2/3} V_f^{1/3} (h^2 d)^{-1/3} w x_e^{2/3} \qquad (2.15)$$

where x_e is the electrode length, h, the cell half-height, d, the width of the cell, and w the electrode width. This equation holds when the flow rate is sufficiently fast that the diffusion layer thickness is much smaller than the depth of the cell. It predicts that i_L varies with the cube root of the volume flow rate.

Unlike the RDE, the channel electrode is said to b e non-uniformly accessible since the current density is greater at the leading edge of the electrode than at the trailing edge. In contrast, the current density is uniform over the surface of a RDE.

The dropping mercury electrode

Historically, the most famous example of a hydrodynamic electrode is the dropping mercury electrode (DME) Heyrovsky (1922). This consists of a reservoir containing pure mercury linked to a fine capillary. The latter is positioned in the electrolyte solution under investigation and mercury allowed to flow slowly from the reservoir and out of the capillary (Fig. 2.15) in the form of drops. These drops act as the electrode. An important advantage is that a fresh electrode surface appears with each new drop. This is helpful when studying species that can 'poison' the surface and prevent electrolysis from proceeding.

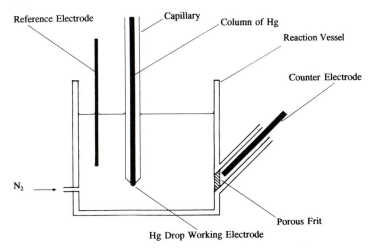

Fig. 2.15 The dropping mercury electrode, D.M.E.

The key feature of hydrodynamic electrodes is that they ensure a steady supply of electroactive material. In the DME this occurs as each drop expands into solution prior to dropping off the capillary and encounters 'fresh' solution which has not been electrolysed. When it leaves the capillary, the solution is stirred so eliminating the products of electrolysis.

Needless to say, the theoretical description of mass transport to the DME is challenging. The electrode has found its major applications for analytical purposes.

2.7   Microelectrodes

The electrodes discussed so far have dimensions in the range of centimetres and millimetres. Microelectrodes typically possess dimensions of around 10pm or below. These tiny electrodes possess some significant advantages over their macroscopic counterparts. An obvious advantage occurs when the voltammetry must be performed in a small volume. For example, microelectrodes have been employed for analytical, *in vivo* electrochemical measurements. The advantages of microelectrodes, however, go far beyond the simple restrictions of space.

The current measured at an electrode is a function of its area. Consequently, the current measured at a microelectrode is significantly lower than that from a conventional macroelectrode (often nano-amps or below). Perhaps surprisingly this presents certain advantages.

First, microelectrodes, by virtue of passing minute currents, induce only a tiny amount of electrolysis in solution. It follows that the diffusion layers of such electrodes will be very thin (of the order of µm in size) and the concentration gradients induced across them will be correspondingly high. Consequently, the rate of mass transport to microelectrodes is *much* greater than for macroelectrodes. This becomes valuable in the study of electrolysis mechanisms when it is wished to study very fast processes, such as rapid rates of electron transfer between electrode and substrate or a fast chemical reaction forming part of the overall mechanism. If the process of interest is fast compared to the rate of mass transport then electrochemical experiments simply reflect the rate of the slowest step in the process and provide information on the rate of diffusion in solution only. With microelectrodes the substantially increased rate of mass transport means that this limitation is less likely to arise and much faster chemical steps are accessible to study by electrochemical methods.

Secondly, as will be revealed in Chapter 3, much useful information about electrode processes can be found by sweeping the electrode potential whilst monitoring the current (linear-sweep or cyclic voltammetry). This effect of changing the electrode potential, in addition to electrolysis, also changes the ionic distribution of electrolyte (background or otherwise) near the electrode surface and this causes a 'charging current' to flow which can mask the phenomena under study. Microelectrodes have a much reduced electrode area which significantly reduces this transient current. It follows that microelectrode systems suffer less distortion from so-called 'capacitative charging', or, alternatively, the rate at which the potential is swept can be made much greater in microelectrode systems. The considerable importance of the latter will be seen in the discussion of cyclic voltammetry in Chapter 3.

Thirdly, it was noted that one benefit of the presence of background electrolyte was the reduction of the bulk solution resistance so as to prevent the latter from masking the study of the current/voltage characteristics of the interface of interest. Physically, this occurs since the passage of current through a resistive solution implies that the electrical potential of the latter is not uniform, as is invariably assumed in the theoretical description of electrode processes (see Chapter 1). Consequently, experimental measurements of resistive solutions lead to distortions of the current/voltage responses from those expected. These are said to be due to 'ohmic drop' and often arise to some small extent, in all macroelectrode experiments. As microelectrodes pass tiny currents this effect is minimized and it is possible to avoid the addition of background electrolyte. Since macroelectrode experiments require background electrolyte this restricts electrochemical investigations with such electrodes to solvents that are polar and therefore capable of dissolving a high concentration of electrolyte. In contrast, microelectrodes permit electrochemical experiments to be performed using non-polar solvents such as benzene or

At first sight it may seem paradoxical that electrodes passing tiny currents should be possess high rates of mass transport. Note however that the latter is quantified by fluxes of electroactive material which measure the amount of material passing in unit time through *unit area*. Thus although the currents are small the current density (current per unit area) is high.

toluene. In addition, microelectrodes have also been used in frozen solvents, supercritical fluids, and even in the gas phase.

References

Brett C.M.A. and Oliveira Brett A.M.C.F. (1986), *Comprehensive Chemical Kinetics,* 26, 355, Elsevier, Amsterdam.

Britz D. (1988), *Digital Simulation in Electrochemistry*, Springer-Verlag, Berlin.

Fick A. (1855), *Philosophical Magazine*, 10, 30.

Fleischmann M., Pons S., Rollison D.R. and Schmidt P.P. (1987), *Ultramicroelectrodes*, Datatech systems, Morganton, NC.

Heyrovsky J., (1922) *Chem. Listy*, 16, 256.

Levich B. (1962), *Physicochemical hydrodynamics*, Prentice-Hall.

Oldham K.B. and Zoski C.G. (1986), *Comprehensive Chemical Kinetics*, 26, 141, Elsevier, Amsterdam.

Wrightman R.M. and Wipf D.O. (1990), *Acc. Chem. Res.* 23, 64.

3 Voltammetry: the study of electrolysis mechanisms

The measurement of electrode currents as a function of the voltage applied to the electrolysis cell can provide remarkably detailed information about the mechanism of the cell reaction of interest. The concepts introduced in Chapter 1 provide the basis for the rationalisation of such voltammetric experiments.

3.1 The experimental measurement of current/voltage characteristics

The simplest approach to the measurement of current/voltage characteristics involves the use of two electrodes. One is termed a working electrode and this is where the reaction of interest occurs. The second is a reference electrode, which should provide a stable and fixed potential so that when a voltage is applied between the two electrodes the drop in potential between the working electrode and the solution, $\phi_m - \phi_s$, is precisely defined. This difference is the driving force for electrolysis to occur at that interface, and the dependence of the current on this quantity was discussed in Chapter 1.

The simple two-electrode arrangement (see Fig. 1.2) is perfectly acceptable for the measurement of current/voltage curves where only a tiny current is passed. For example, it is ideal for microelectrode studies. However, for larger electrodes a difficulty arises. Consider a voltage, E, applied between a large working electrode and a reference electrode and assume that a finite current is flowing between them. Then,

$$E = (\phi_m - \phi_s) + iR + (\phi_s - \phi_{REF}) \tag{3.1}$$

Notice that E is split into three terms. The first of these, $(\phi_m - \phi_s)$, is the driving force for electrolysis at the working electrode/solution interface. The second describes the voltage drop, iR, in solution due to passage of current i between the two electrodes. R is the electrical resistance of the intervening solution. The third term, $(\phi_s - \phi_{REF})$ is the potential drop at the reference electrode/solution interface and is fixed by the chemical composition of the chosen reference electrode.

The aim of any voltammetry experiment is to measure i as a function of *changes* in the quantity $(\phi_m - \phi_s)$. For microelectrode experiments the term iR can be neglected and since $(\phi_s - \phi_{REF})$ is fixed, eqn 3.1 simply reduces to

$$E = (\phi_m - \phi_s) + \text{constant}$$

Consequently, changes in the applied potential E are directly reflected in the driving force, $(\phi_m - \phi_s)$, as desired, and voltammetry is meaningful with two electrodes. With large electrodes, however, iR is no longer negligible and so changes in the applied potential are not confined to changes in $(\phi_m - \phi_s)$. In particular they alter the iR term since the current flowing through the cell will also be induced to change. Moreover, the passage of large currents through the reference electrode can change its chemical composition and so the third term, $(\phi_s - \phi_{REF})$, may no longer be constant. It can be concluded that two-electrode experiments with other than tiny currents will be impossible to interpret using the relations established in Chapter 1 since the current will not be readily related to the potential drop at the working electrode/solution interface $(\phi_m - \phi_s)$.

To circumvent this problem it is usual to conduct voltammetric experiments using a three-electrode system in which an auxiliary or counter electrode is used, in addition to the other two. These are controlled by electronics (a potentiostat) which ensures that current only flows between the working and auxilliary electrode. The potential of the working electrode is held relative to the now stable reference electrode and the electronics ensure that no current passes through the reference arm of the circuit. Figure 3.1 shows a typical three-electrode arrangement for an RDE experiment.

The RDE is 'jacketted' by water at a controlled temperature to ensure the experimental solution is maintained at a constant and known temperature.

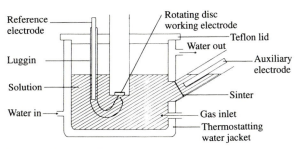

Fig. 3.1 Schematic of a typical rotating disc experiment.

In a 'typical' RDE experiment the solution is out-gassed with an 'inert' gas such as nitrogen to remove oxygen from the electrolyte solution. This is necessary since oxygen is a good scavenger of radicals which may be produced during the course of the electrode reaction.

The working electrode is held in the main chamber along with the reference electrode. The auxilliary electrode is held in a separate compartment separated by a frit as electrolysis also occurs at this electrode. The frit prevents any products formed at the auxiliary electrode from entering the main part of the cell.

3.2 Voltammetry

In this section different voltammetric experiments are introduced. It should be assumed throughout that the driving force $(\phi_m - \phi_s)$ is controlled as above, either through the use of a micro-working electrode or with a three-electrode potentiostat. Also, for simplicity, assume that the electrolyte solution contains an electroactive species A which can undergo a simple one-electron (n = 1) reduction to form B

$$A(aq) + e^-(m) \xrightleftharpoons[k_{ox}]{k_{red}} B(aq)$$

In Chapter 1 the terms 'reversible' and 'irreversible' were used to characterize two extremes of behaviour of the electrode kinetics of the A/B couple. It will emerge that different voltammetric behaviour is associated with each of these two cases.

Linear-sweep voltammetry

Linear-sweep voltammetry (or its close relative, cyclic voltammetry, discussed below) is invariably the first technique employed in any electrochemical investigation since it is easy to perform and provides quick and useful information about the system under investigation. The experiment is conducted in stationary solution and so relies only on diffusion to transport material to the electrode surface. The mass transport of the species A to the electrode can be predicted using the equation

$$\frac{\partial[A]}{\partial t} = D_A \frac{\partial^2[A]}{\partial x^2}$$

The potential of the working electrode is swept from a value E_1 at which A cannot undergo reduction, to a potential E_2, where the electron transfer is driven rapidly. The applied potential E is a function of the speed at which the potential is swept (v_s) and the time of the sweep (t)

$$E(t) = E_1 - v_s t$$

Figure 3.2 shows the form of the potential ramp and the corresponding current response. For macroelectrodes the electrode potential is typically swept at rates in the range 5 mV s^{-1} to more than 10^2 V s^{-1}.

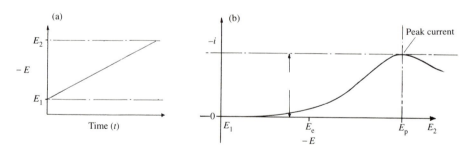

Fig. 3.2 The potential sweep (a) and corresponding current response (b) for an irreversible electron transfer reaction.

The form of the current/voltage behaviour can be understood as follows, on the initial assumption that the A/B couple has irreversible electrode kinetics. Initially, no current is passed since the applied potential is not great enough to induce electron transfer. But as the potential is swept to more reducing potentials it reaches values that are capable of inducing the reduction of A to B at the electrode, and current starts to pass. As E (and so $\phi_m - \phi_s$) is made more negative, the electrochemical rate constant, k_{red}, for the reduction of A becomes greater as suggested by eqn 1.21 and, initially, the current rises approximately exponentially with potential (or time). As even more negative potentials are reached, the rise becomes less than exponential and ultimately a maximum is reached, after which the current falls off.

At first sight the occurrence of a maximum appears paradoxical since eqn 1.21 predicts k_{red} should increase steadily with E. However, the current flowing reflects not only k_{red} but also the surface concentration, $[A]_o$, of the reactant eqn 1.7. As the potential is swept to more negative valuesalthough k_{red} increases, $[A]_o$ steadily decreases as electrolysis consumes A, which is only partially replenished by diffusion of fresh A from bulk solution. The maximum in the current/voltage curve then reflects a balance between an increasing heterogeneous rate constant and a decrease in surface concentration. The maximum current is known as the *peak current*, i_p. Once the peak current is attained the magnitude of the current flowing is simply controlled by the rate at which fresh A can diffuse up to the electrode surface. The fall in current arises since, as electrolysis proceeds, the zone around the electrode in which A is depleted becomes thicker and so new A has further to diffuse to the electrode. Algebraically, the fall in current can be accounted for in terms of eqn 2.4 and a steadily increasing diffusion layer thickness resulting from progressive electrolysis of the solution. Thus the part of the linear-sweep voltammogram at potentials more negative than that corresponding to i_p simply reflects the rate at which A can diffuse in solution, whereas the part of the current/voltage curve that precedes the peak is controlled by the electrode kinetics.

Figure 3.3(a) shows a voltammogram for the case where the A/B couple now has reversible electrode kinetics. Also shown is the irreversible voltammogram previously discussed. Notice that in the reversible case appreciable currents flow when the electrode attains a potential more negative than E_e. In other words, as soon as the reduction becomes thermodynamically viable it takes place as a result of the rapid electrode kinetics implicit for a reversible couple. In contrast no currents, flow in the irreversible case until the potential

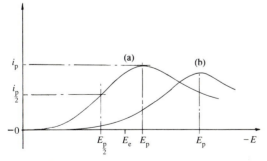

Fig. 3.3 Linear sweep voltammogram for (a) a reversible electron transfer and (b) an irreversible electrode reaction.

is considerably more negative than E_e. In the terminology of Chapter 1 an overpotential ('over' and above the thermodynamic potential) has to be applied to drive the reduction of A to B. Hence E_p occurs at more negative potentials in irreversible systems.

Two other important points of comparison between reversible and irreversible voltammograms emerge:

(a) In the reversible case, the peak potential, E_p, is constant and independent of sweep rate, whereas for irreversible reductions systems E_p shifts to more negative potentials at faster sweep rates. Quantitatively, for the reversible process

$$\left| E_p - E_{\frac{p}{2}} \right| = 2.20 \frac{RT}{F}$$

whereas for the irreversible electron transfer

$\left| E_p \right|$ shifts by approximately $1.16 \dfrac{RT}{\alpha F}$ for each factor of 10

$$\left| E_p - E_{\frac{p}{2}} \right| = 1.86 \frac{RT}{\alpha F}$$

(b) The peak current is larger for a reversible couple than for an irreversible couple for the same voltage sweep rate. Quantitatively, for the reversible case

$$\left| i_p \right| = 0.4663^{\frac{3}{2}} AFD_A^{\frac{1}{2}} v_s^{\frac{1}{2}} [A]_{bulk} \left(\frac{F}{RT} \right)^{\frac{1}{2}}$$

whereas for the irreversible reaction

$$\left| i_p \right| = 2.99 \times 10^5 (\alpha)^{\frac{1}{2}} AD_A^{\frac{1}{2}} v_s^{\frac{1}{2}} [A]_{bulk}$$

The above differences reflect the contrasting electrode kinetics of reversible and irreversible systems. In the case of a reversible system the electrode kinetics are such that for potentials near E_e both k_{red} and k_{ox} are large. In fact as a result of this the concentrations of A and B *at the electrode surface* are effectively in equilibrium with their ratio predicted by the Nernst equation. In contrast, electrolysis in irreversible systems proceeds with negligible back reaction. However, for any system, reversible or not, i_p is directly proportional to the concentration of A and increases with voltage sweep rate. Figure 3.4 illustrates this dependence for the case of a reversible system, and i_p varies as the square root of the sweep rate.

The increase of i_p with v_s may be interpreted as follows. Recall that i_p shows the flux of material reacting at the electrode surface and that near E_p this is controlled by the rate of diffusion of A. Specifically, Fick's first law eqn 2.1 indicates that i_p reflects the concentration gradient of A near the electrode, which in turn is controlled by the diffusion layer thickness. If the

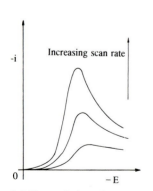

Fig. 3.4 The variation of current with voltage scan rate for a reversible electron transfer reaction.

electrode potential is swept more rapidly, relatively less time is available for electrolysis and the depletion of A near the electrode is reduced, resulting in a thinner diffusion layer and hence a steeper concentration gradient. The resulting larger flux gives rise to an enhanced i_p.

Cyclic voltammetry

Linear-sweep voltammetry can be extended so that when the potential reaches the value E_2 the direction of sweep is reversed and the electrode potential is scanned back to the original value, E_1 Fig. 3.5. This gives a triangular potential cycle and the technique is termed cyclic voltammetry.

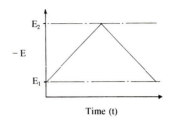

The shape of the voltammogram on the forward sweep is identical to that described above for the case of linear-sweep voltammetry and depends on the reversibility of the A/B redox couple. However, on reaching E_2 the potential is then swept back, oxidizing species B, formed at the electrode during the forward scan to species A. A current in the opposite sense to the forward scan, is observed due to oxidation of B to A. This current increases initially since a high concentration of B is present in the diffusion layer and the kinetics for the conversion of B to A become more favourable as the potential becomes more positive. Gradually, all of B present in the diffusion layer is reconverted to A and the current drops to zero.

Fig. 3.5 Variation of the applied potential as a function of time in a cyclic voltammetry experiment.

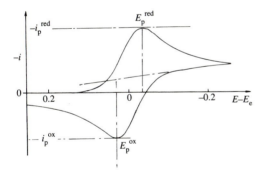

Fig. 3.6 Cyclic voltammogram for a reversible electron transfer reaction.

The peak potential and the peak size seen on the reverse scan again reflect the reversibility of the A/B couple. Figure 3.6 shows a cyclic voltammogram for a reversible couple and significant oxidative currents flow at potentials immediately anodic of E_e. In this case, the heights of the forward and reverse current peaks are of the same magnitude and are separated by a potential of approximately 59 mV (at 25 °C) which is independent of scan rate. If n electrons are transferred in a reversible electrode process the separation becomes

$$\left| E_p^{ox} - E_p^{red} \right| = 2.218 \frac{RT}{nF}$$

Figure 3.7a shows the cyclic voltammetric behaviour for an irreversible electrode couple. Notice that for B to be reconverted into A an appreciable overpotential is required and the reverse peak only appears at potentials very much more negative than E_e. The size of the reverse peak relative to the forward peak will depend on the voltage sweep rate. As for linear-sweep voltammetry, the potentials of both the forward and reverse peaks are voltage sweep-rate dependent and this contrast with the reversible case is diagnostic of the nature of the electrode kinetics.

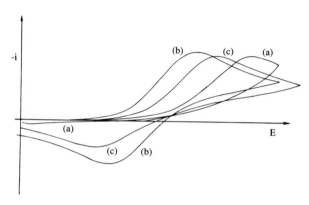

Fig. 3.8 Variation of i_p as a function of voltage scan rate.

Fig. 3.7 Cyclic voltammogram for (a) an irreversible, (b) a reversible, and (c) a quasi-irreversible electron transfer reaction.

Regardless of reversibility, the absolute magnitudes of the peak currents for both forward and reverse scans depend on the voltage sweep rate. Figure 3.8 shows this dependence for a reversible system where the ratio of the forward and reverse peak currents is unity, but i_p for both peaks varies with $v_s^{1/2}$. The explanation is essentially the same as given in the preceding section.

Note that the terms 'reversible' and 'irreversible' refer to limiting cases, according to whether the electrode kinetics are fast or slow relative to the mass transport conditions of the electrode of interest. Intermediate cases (quasi-reversible behaviour) exist. The surface concentrations of A and B now depend on both the forward and reverse electron transfer rates and the rate of mass transport; Fig. 3.7 also shows the behaviour of a quasi-reversible electrochemical reaction. A reverse peak of similar magnitude to the forward peak is observed on the reverse sweep, but, unlike the case of reversible kinetics the separation of the two peaks is dependent on the scan rate and the peak current is not proportional to the square root of the scan rate. Only a tiny peak will be seen unless a very fast scan rate is used because the electrode has to be taken to very oxidising potentials so as to 'drive' the B→A reaction. Otherwise, almost all the B formed in the forward scan will diffuse into bulk solution and be unavailable for re-oxidation.

Potential step chronoamperometry

The preceding section identified the voltage scan rate, v_s, as the key parameter in cyclic or linear-sweep voltammetry. A limiting case of the latter experiment occured where the potential of the working electrode is instantaneously stepped between the values E_1 and E_2 corresponding, respectively, to no electrolysis and to complete conversion of A to B at the electrode surface, as in Fig. 3.9.

Immediately following the step, a large current is detected which falls steadily with time. This arises since the magnitude of the current is controlled by the rate of diffusion of A to the electrode. The concentration gradients shortly after the step are extremely large since there has been little time for any depletion of the electroactive material. Consequently, the currents flowing are very large initially. Gradually, as depletion occurs, the diffusion layer thickness increases and the current decreases : ultimately to zero, as shown in Fig. 3.10. The Cottrell equation describes the current response as function of time

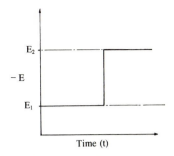

Fig. 3.9 The variation of applied potential in a potential step experiment.

$$|i| = \frac{nFAD_A^{\frac{1}{2}}[A]_{bulk}}{\pi^{\frac{1}{2}}t^{\frac{1}{2}}} \qquad (3.2)$$

This equation suggests that potential step experiments can be used to measure diffusion coefficients. *No* information about the electrode kinetics emerges provided E_2 is constrained to potentials negative (for a reduction) of the cyclic voltammetric peak current.

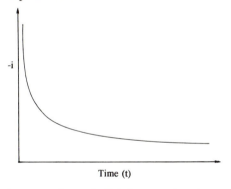

Fig. 3.10 The current response in a potential step experiment.

3.3 Current-voltage curves in the prescence of coupled convection and diffusional transport

In Section 2.6 the merits of employing hydrodynamic electrodes were introduced. The voltammograms resulting from such experiments are considered next and a typical rotating disc current/voltage curve for the reduction of A to B is shown in Fig. 3.11 In normal practice such voltammograms are recorded using slow voltage scan rates (less than 5 mV s^{-1} is typical). Under these conditions no peak is seen, in contrast to the cyclic voltammetry experiment. This can be rationalized again by considering a potential scan between the limits E_1 and E_2 previously considered. Initially, the electrode potential is insufficient to induce the reduction of A. As it is made more negative, electrolysis starts and the current rises, approximately exponentially as in cyclic voltammetry. For more negative potentials the current rises to a maximum value (the transport-limited current). This current is maintained at subsequent negative potentials because the presence of a convective flow supplements the supply of electroactive A from bulk solution to the electrode surface. This constant replenishment ensures that a steady current is maintained. The magnitude of this is controlled by the extent of convective transport to the electrode (the term transport-limited current) and depends on the speed at which the electrode is rotated, for the case of the rotating disc electrode. Figure 3.11 shows a series of voltammograms recorded for a reversible electron transfer reaction at different rotation speeds. The limiting currents can be analysed as described in Section 2.5 (p. 17). The rising part of the current/voltage curves in Fig. 3.11 reflects the electrode kinetics of the A/B couple and suitable analysis permits the reversibility, or otherwise, of the couple to be inferred.

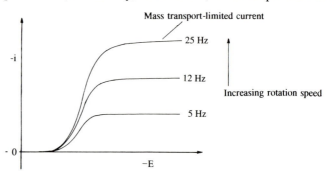

Fig. 3.11 Current/voltage curves recorded at an RDE for a range of rotation speeds.

3.4 Electrolysis reaction mechanisms

In the discussion of voltammetry so far it has been assumed that the product of the electrode process (B) is chemically stable. In reality most electrode reactions of interest generate species that are highly reactive and are therefore capable of undergoing further reaction at the electrode or in solution, creating a complex electrolysis mechanism. The aim of this section is to illustrate how voltammetric investigations can reveal such mechanisms.

A huge range of electrolysis mechanisms are conceivable. For example consider the reduction of a simple alkyl halide:

$$RX(aq) + e^-(m) \longrightarrow R(aq) + X^-(aq)$$

The addition of an electron is thought to induce the rapid cleavage of the carbon–halogen bond with the formation of a halide anion and an alkyl radical. The latter might be adsorbed on to the electrode surface or released into solution. In either case subsequent reaction may occur as shown in Fig. 3.12 which defines a range of possible products resulting from electrolysis of RX.

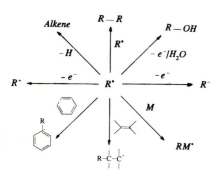

Fig. 3.12 Possible fates of the electrogenerated radical R.

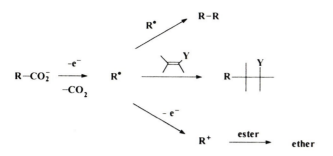

The idea that electrode reactions may occur via complex mechanistic processes is not new. Faraday noted in some of his original electrolysis experiments that hydrocarbon formation occurs when electrolysing aqueous solutions of acetate. In 1849 Kolbe followed up these original observations and proposed a reaction mechanism to account for the observations:

By suitable selection of the electrode potential, current density, solvent, supporting electrolyte, solution pH, temperature, deliberate addition of other reagents, etc., the selective formation of one product may be favoured. The topic of electrosynthesis is considered in Section 3.4 (p. 34) after some mechanistic considerations are examined.

The scope for considerable mechanistic complexity is clearly evident from the above. In the following discussionvarious relatively simple systems, both organic and inorganic, are shown to be amenable to mechanistic interrogation using solely voltammetric techniques. It should be noted that the study of more complex processes has led to the adaptation of a diverse range of spectroscopic tools to study electrochemical reactions. A description of some of the techniques available is given in Chapter 5.

Coupled homogeneous reactions

The use of electrochemical techniques for the investigation of simple electron transfer reactions has already been identified in Chapters 1 and 2. In this section the analysis of voltammetric behaviour examining electron transfer reactions that are coupled to homogeneous reactions is described.

A notation adopted to describe the reaction sequence of an electrolysis mechanism has been described by Testa and Reinmuth (1961). A heterogeneous electron transfer at the electrode is denoted as an 'E' step and a homogeneous chemical reaction is termed a 'C' step. For example

$$\text{E} \qquad A(aq) + e^-(m) \rightleftharpoons B(aq)$$

$$\text{C} \qquad B(aq) \xrightarrow{k_{EC}} \text{products}$$

would be referred to as an EC reaction, since the electron transfer (E) is followed by a first-order homogeneous decay (C) of the electrochemically generated species (B).

To establish a quantitative knowledge of such reactions the mass transport equations described in Section 2.1 for the movement of chemical species must be modfied to account for the loss of a species in solution by chemical reaction. For the EC reaction under diffusional mass transport (in one dimension), the equations for species A and B are

$$\frac{\partial [A]}{\partial t} = D_A \frac{\partial^2 [A]}{\partial x^2} \qquad (3.1)$$

$$\frac{\partial[B]}{\partial t} = D_B \frac{\partial^2[B]}{\partial x^2} - k_{EC}[B] \qquad (3.2)$$

Equation 3.2, which describes the mass transport of species A, is identical to that for a chemically stable species since it is not lost by any homogeneous reaction. The diffusion equation for species B, however, must be modified as species B now undergoes a homogeneous first-order decay at a rate $k_{EC}[B]$. The diffusion equation contains a term on the right-hand side that quantifies the loss of species B.

Typically, cyclic voltammetry is the first experimental technique used when investigating the reactivity of a new molecule. The technique is quick and easy to perform and establishes the basic voltammetric response of the system. If the species generated are unstable then the current/voltage characteristics differ from those described for the case of a stable, reversible electron transfer (Section 3.2). Some examples of simple electrode mechanisms are now considered.

The EC reaction

For the EC reaction species B, formed at the electrode, is unstable and decays with a rate constant k_{EC}. If k_{EC} is significantly fast then as soon as B is formed at the electrode it is lost through homogeneous reaction. The cyclic voltammetry observed in this situation differs from that of a stable electrode reaction. The forward voltammetric sweep reducing A to B, is similar to that detected for the reversible, stable cyclic voltammogram. However, as the potential is swept back from E_2 to E_1 the reverse peak due to the re-oxidation of B to A is lost, since B has been destroyed through homogeneous reaction (Fig. 3.13). This is now a chemically irreversible system (as opposed to the irreversible electron transfer noted in Chapter 2 which occurs because a high enough overpotential cannot be reached to drive species B back to A electrochemically).

For the situation where k_{EC} is very slow, then little or no difference from the stable cyclic voltammogram is detected (Fig. 3.6) as B is not significantly depleted by reaction. Intermediate reaction rates result in cyclic voltammograms where the height of the reverse peak varies reflecting the stability of species B (Fig. 3.13). The relative heights of the back peaks can then be used to estimate the rate of reaction in solution.

The analysis of the reaction kinetics must, however, account for the influence of the potential sweep rate. For example for a reaction that has an intermediate k_{EC} rate as the scan rate is increased the time for the reactive species to decay is reduced and a larger back peak is detected on the reverse sweep. The concept of time-scale is therefore an important aspect of an electrochemical investigation and, in particular, the successful analysis of data from cyclic voltammetric measurements must account for this. From Fig. 3.13 it is clear that not only does the relative height of the back peak depend upon scan rate, so does the position of the reduction wave on the forward sweep shifts.

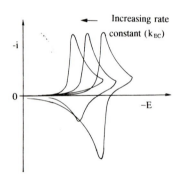

Fig. 3.13 Cyclic voltammograms showing the influence of chemical reaction on the observed voltammetric behaviour.

In particular, as the rate constant increases the wave shifts anodically and it becomes easier to reduce species A. This can be rationalized by recalling the Nernst equation (eqn 1.3), which predicts the ratio of equilbrium surface concentrations of A and B at a particular potential for an electrochemically reversible electrode reaction. The homogeneous reaction removing B perturbs the surface equilibrium, and, in order to try and restablish the equilibrium concentration at the electrode surface, the reaction is driven towards the right-hand side, producing more B. This replenishes some B lost by reaction and causes an anodic shift in the wave.

An example of an EC reaction is the isomerization of *fac*-Mn(dpe)$_2$(CO)$_3$Cl. In Fig. 3.14a a cyclic voltammogram for the oxidation of *fac*-Mn(dpe)$_2$(CO)$_3$Cl is shown, the reverse sweep showing only a small reverse peak. This was rationalized by Bond (1977) who proposed that this behaviour was the result of fast isomerisation of *fac*-Mn(dpe)$_2$(CO)$_3$Cl$^+$ to the *mer*-isomer (Fig. 3.15). More information was available from this particular study, since *mer*-Mn(dpe)$_2$(CO)$_3$Cl$^+$ exhibits reversible electrochemical behaviour at a lower potential to the *fac*-isomer. Figure 3.14(b) shows the extended voltammetric behaviour for two successive sweeps. The first scan shows just one oxidation corresponding to the one-electron oxidation to *fac*-Mn(dpe)$_2$(CO)$_3$Cl$^+$. However, the reverse sweep at lower potentials shows a new feature due to the reduction of the *mer*-cation product. On the second sweep, a new oxidation feature from the *mer*-isomer is detected. Analysis of the peaks heights as a function of the scan rate enabled a quantitative estimate of the rate of the isomerization reaction.

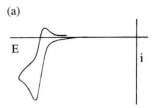

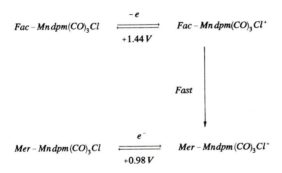

Fig. 3.15 The proposed reaction sequence.

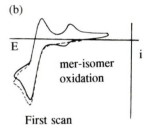

Fig. 3.14 Cyclic voltammograms of the manganese *mer/fac* system for (a) the oxidation of the *mer*-isomer and (b) an extended voltage range and the additional voltage sweep. As reported by Bond (1977).

The ECE reaction

An extension of the EC reaction is the ECE process

E $A(aq) + e^-(m) \rightleftharpoons B(aq)$

C $B(aq) \xrightarrow{k_{ECE}} C(aq)$

E $C(aq) + e^-(m) \rightleftharpoons D(aq)$

The initial electrochemical and homogeneous chemical steps are identical to the EC reaction. However, the product formed by the homogeneous chemical reaction C is now also electroactive. The reduction of the new product C to D usually occurs at a different potential (either higher or lower) than that of the A to B process. The case where C is more difficult to reduce than A is described with reference to cyclic voltametry measurements. The situation where C is more easily reduced than A is described by reference to coupled convection–diffusion measurements.

When A is more easily reduced than C then the cyclic voltammogram recorded has two separate electrochemical features (Fig. 3.16). On the reductive sweep a peak is detected due to the simple A to B electrode reaction, then, at more negative potentials a further voltammetric wave is observed (the magnitude of which is dependent upon the rate constant kECE) due to the reduction of C to D. When the sweep is reversed two peaks are detected (again the magnitude of the detected current depends upon the rate constant k_{ECE}). The first corresponds to the oxidation of D to C and the second to the oxidation of B to A (Fig. 3.16). The relative heights of the observed peaks is a function of the homogeneous kinetics and potential scan rate. Analysis of the relative peak heights provides information about the magnitude of the rate constant k_{ECE}.

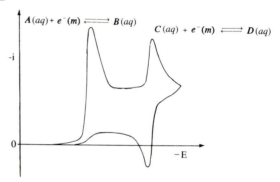

Fig. 3.16 Cyclic voltammogram of ECE reaction.

Further information about ECE reactions can be obtained from the use of hydrodynamic voltammetry, such as RDE experiments. In particular, the ability to control the mass transport rate by simply altering the rotation speed enables the reaction to be probed over a wide range of mass transport rates.

The current/voltage characteristics for an ECE mechanism where the reduction of C is easier than that of A, is shown in Fig. 3.17. The enhancement of current over the simple one-electron occurs since the current voltage curve now contains the current for both the A to B and the C to D reductions.

Analysis of the steady state experimental data is performed in terms of the number of effective electrons transfered (N_{eff}) where

$$N_{eff} = \frac{\text{Total current of process}}{\text{Current for A} \longrightarrow \text{B process alone}}$$

The N_{eff} is measured at the rotating disc electrode by variation of the rotation speed and monitoring the limiting current.

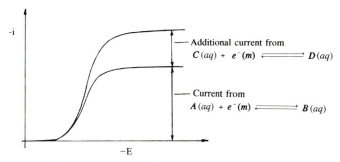

Fig. 3.17 RDE voltammogram for an ECE process.

The measured N_{eff} is dependent on both the homogeneous rate constant and the rotation speed. The influence of the rotation speed occurs as convection draws reactant to the electrode surface and then sweeps away products from the electrode reaction. For a current to be detected from the reductive process of C to D, B must decay faster than it is lost by the mass transport from the electrode interface. This effect can be best clarified by considering the limits of kinetic behaviour at a single rotation speed. First, if k_{ECE} is small, then B is removed from the electrode before it can decay to C, and so no contribution from the reduction of C to D is detected in the current voltage curve; N_{eff} is therefore 1. If k_{ECE} is very large, then as soon as B is formed electrochemically it is converted almost instantaneously to species C which can then be further reduced at the electrode to D, before mass transport can remove it from the electrode. The transport-limited current detected at E_2 therefore corresponds to a two electron process, i.e. the reduction of A to B and then C to D, and N_{eff} is therefore 2. For intermediate rate constants only partial conversion of B to C occurs before removal of species B from the electrode surface and the measured N_{eff} drops below the possible maximum of 2.

This argument can be repeated for all possible rotation speeds and a 'working curve' describing the variation of N_{eff} as a function of the rotation speed centre drawn, (Fig. 3.18). This working curve may be used to examine experimental data and gain a value for the rate constant K_{ECE}.

An example of an ECE mechanism is the reduction of nitroethane, which, after reduction, undergoes a homogeneous reaction with H^+. The product can then undergo a further reduction to yield the corresponding nitroso compound.

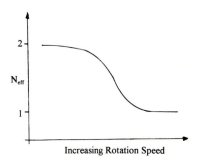

Fig. 3.18 The variation of N_{eff} as a function of rotation speed.

$$CH_3 - CH_2 - NO_2(aq) + e^-(m) \rightleftharpoons (CH_3 - CH_2 - NO_2)^{\cdot -} \ (aq)$$

$$H^+(aq) + (CH_3 - CH_2NO_2)H^{\cdot -} \ (aq) \longrightarrow (CH_3 - CH_2 - NO_2 . H)(aq)$$

$$(CH_3 - CH_2 - NO_2H)(aq) + e^-(m) \rightleftharpoons CH_3 - CH_2 - NO(aq) + OH^-(aq)$$

The EC′ Mechanism

The final process to be considered in detail is the EC′ reaction, where the prime (′) represents a catalytic process. Catalytic reactions have received considerable attention since they often represent a clean and efficient method of enhancing chemical reactivity. The advantages of catalytic processes make them particularly useful industrially, where they can produce considerable cost improvements. The EC' mechanism can be summarized as

$$A(aq) + e^-(m) \rightleftharpoons B(aq)$$

$$B(aq) + Y(aq) \longrightarrow A(aq) + \text{products}$$

The catalytic process occurs as the product of the electrode reaction, B, reacts with a substrate molecule Y in solution. The result of this reaction is that B is oxidized back to the starting material A (the catalytic cycle) and Y is converted to products.

Figure 3.19 shows a typical cyclic voltammogram for differing quantities of substrate Y, the dashed line shows the behaviour when no substrate is present. An enhancement in the current is detected as the homogeneous reaction regenerates A, which can then be reduced again to B, at the electrode. The size of the catalytic current detected is dependent on the quantity of substrate present in solution and the rate of the reaction between B and Y.

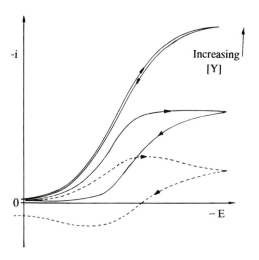

Fig. 3.19 Cyclic voltammogram for EC′ process

Voltammograms recorded at the RDE also show a catalytic enhancement in current. Analysis of the reaction again occurs by interpretation of N_{eff} meaurements as a function of the rotation speed. Unlike the ECE mechanism, however, it is not possible to generate a universal working curve describing the relationship between N_{eff} and rotation speed. This is because the size of the enhancement for any rotation speed is dependent upon the ratio of [A]/[Y]. Consequently, analysis of EC′ mechanisms is either performed so that the substrate is in large excess and the reaction acts in a pesudo-first-order manner, or computational models are employed to predict the particular behaviour for a given [A]/[Y] ratio.

In the above sections just a few simple examples of possible electrolysis mechanisms have been discussed. There are, however, a huge range of other possible schemes that may occur. For example, an electrochemically inactive material may undergo homogeneous reaction forming a product that is electroactive. This type of process is termed a CE reaction and can be described by the following reaction scheme

$$O(aq) \rightleftharpoons A(aq)$$

$$A(aq) + e^-(m) \rightleftharpoons B(aq)$$

Further complex mechanisms can be drawn, involving second-order steps such as disproportionation, or the EC_2 process where species B decays via a bimolecular reaction:

$$A(aq) + e^-(m) \rightleftharpoons B(aq)$$

$$B(aq) + B(aq) \longrightarrow \text{products}$$

More esoteric still are square sequences where all the species are electrochemically active and equilibrium exists between various components in the system

$$
\begin{array}{ccc}
A(aq) + e^-(m) & \rightleftharpoons & A^-(aq) \\
\Updownarrow & & \Updownarrow \\
B(aq) + e^-(m) & \rightleftharpoons & B^-(aq)
\end{array}
$$

Most of these complex processes can now be analysed using relatively simple computational simulations or analytical theory.

Processes such as the ECE mechanism show how multiple electron transfers can occur through homogeneous reaction. The next section describes a slightly different situation, where again a multiple electron transfer occurs but were the reactant is able to change oxidation state by more than one. This means that no homogeneous reaction is required to provide a pathway for more than one electron transfer.

Multiple electron transfer reactions

The ability to alter the oxidation state of a molecule by more than one is most often encountered in organic electrochemistry and the electrochemistry of inorganic metals (e.g. Tl^+/Tl^{3+}). The simplest multiple electron transfer reaction is the two-electron transfer (EE) reaction. Conceptually, the EE process can occur in two ways, either from a one single step involving both electrons or as the result of two consecutive one-electron steps

The consecutive pathway is the route most commonly observed in multiple electron transfer processes. Simplistically, this may be rationalized, as it is easier to overcome two small activation barriers than one large one.

$$A(aq) + 2e^-(m) \longrightarrow C(aq)$$

$$A(aq) + e^-(m) \longrightarrow B(aq) \quad (E = E_1^0)$$

$$B(aq) + e^-(m) \longrightarrow C(aq) \quad (E = E_2^0)$$

Depending upon the relative values of E_1^0 and E_2^0 the electron transfers can occur at the same or different potentials. Figure 3.20 shows the cyclic voltammogram of a molecule that is capable of undergoing two consecutive, reversible, electron transfer steps, separated by a potential value ΔE.

The situation becomes more complex, however, if $E_2^0 > E_1^0$. Then, the intermediate (B) is thermodynamically unstable. The voltammetric response observed is then dependent upon which of the electron transfer steps is rate determining. In a similar manner to that noted in Chapter 1 an equation can be defined to predict the current response as a function of the overpotential (Bard 1980).

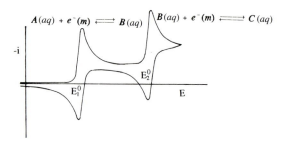

Fig. 3.20 Cyclic voltammogram for a two-electron transfer reaction.

Electrosynthesis

The discussion of electrolysis mechanisms has already shown that a range of complex electrode reactions is possible. In particular, the ability of electrochemical processes to create highly reactive and unusual chemical species in an efficient, selective, and well-controlled manner offers many synthetic possibilities. In the laboratory electrosysnthesis experiments are performed in cells not unlike those described earlier for the RDE (see Fig. 2.8). Although a larger electrode is often used and is solution agitation increased by stirring. Industrially, with the significantly larger quantities involved, the electrode is usually employed in the form of a packed bed which has a high surface area. The solution can access this electrode bed easily and this enables efficient rates of conversion.

The use of electrochemistry as a synthetic tool was appreciated from the early days of electrochemical studies and before the 1920s electrochemistry had been employed to perform some difficult functional group interconversions. As time has progressed techniques have been extended and now electrochemical reactions have become an accepted, if not fully exploited, synthetic tool. Electrosynthesis has been used in both oxidative and reductive modes to bring about a wide variety of synthetic processes. For example

Substitution: the cyanation of hydrocarbons

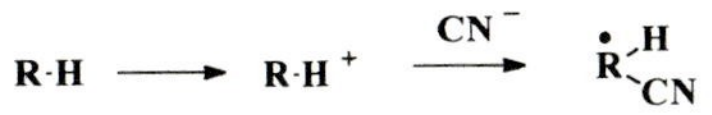

$$R \cdot H \longrightarrow R \cdot H^+ \xrightarrow{CN^-} \overset{\bullet}{R}\overset{H}{\underset{CN}{\diagdown}}$$

Addition: the acetoxylation of stilbene

$$Ph\ CH{=}CH\ Ph \xrightarrow{-e^-} Ph\ \overset{+}{CH}{-}\overset{\bullet}{CH}\ Ph \xrightarrow{2\ AcO^-} \underset{Ph\ \ \ OAc}{\overset{H\ \ \ OAc}{C{-}C}}\overset{Ph}{\underset{H}{\diagdown}}$$

The acetoxylation process also has stereochemical control. It is proposed that the reaction is a concerted process with simultaneous electron transfer to the anode and attack by the AcO⁻ ion.

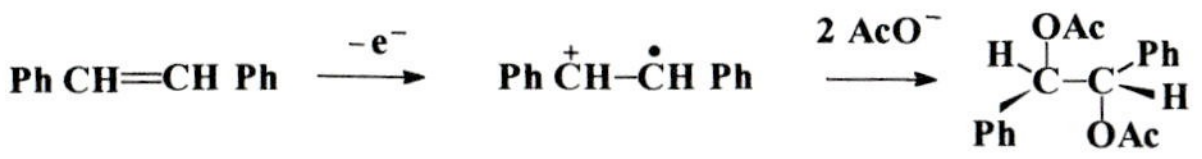

Elimination

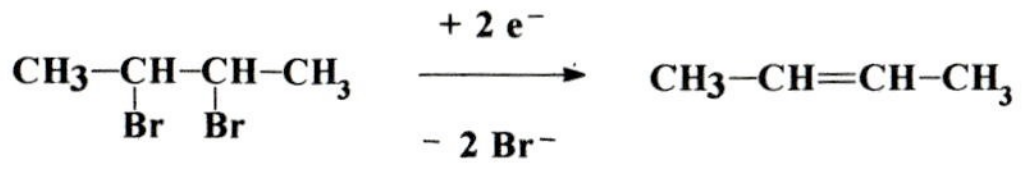

$$\underset{\overset{|}{Br}\ \ \overset{|}{Br}}{CH_3{-}CH{-}CH{-}CH_3} \xrightarrow[-\ 2\ Br^-]{+\ 2\ e^-} CH_3{-}CH{=}CH{-}CH_3$$

Coupling

$$2\ RCO_2^- \xrightarrow{-\ 2\ e^-} 2\ R^{\bullet}\ +\ 2CO_2$$
$$\searrow$$
$$R{-}R$$

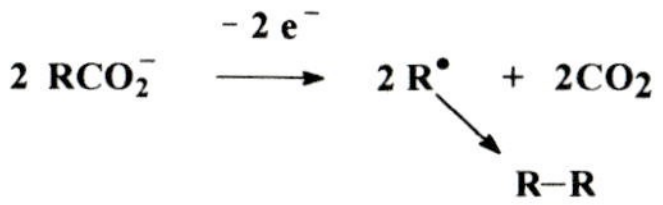

Synthetic routes have increasingly employed such processes and, in particular, the radical coupling reaction has few chemical analogues. Perhaps the greatest use of electrosynthesis, however, is the ability to selectively induce complex chemical reactions not realizable from simple chemical processes, e.g. the synthesis of 8-oxotetrahydropalmatine

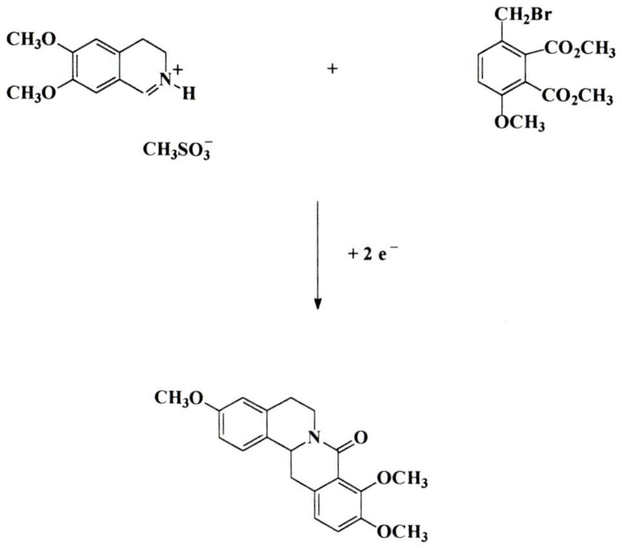

Heterogeneous reactions

The chemical reactions involved in electrolysis meachanisms have so far been confined to the homogeneous phase. However a significant range of heterogeneous coupled chemical reactions can occur. The type of surface processes that occur are electrode/electrolyte interactions (these are considered in Chapter 4) and substrate adsorption/desorption, which can also be coupled to mass transport.

In the following sections the voltammetric response of adsorbed material is discussed and this is followed by a description of some applications of the investigations of such processes.

Surface-adsorbed species

Consider the voltammetric behaviour of a molecule uniformly adsorbed on to an electrode surface, which is capable of undergoing a reversible one-electron transfer. The resulting cyclic voltammogram shows two peaks (Fig. 3.21). On the forward (reductive) scan a peak is observed corresponding to the reduction of adsorbed A to form B, which remains adsorbed to the surface. On the reverse sweep a current in the opposite sense is observed and adsorbed B is reconverted to adsorbed A. Several differences are apparent compared to the solution-phase voltammogram for a stable, reversible, one-electron transfer (see Fig. 3.6).

First, both forward and reverse peaks are now symmetrical and the current drops back to zero at potentials significantly greater than E1. The absence of a current at potentials greater than the reduction potential of A occurs simply because only a limited quantity of A exists on the surface. Once this has been converted to B no further electron transfer can occur. (For the solution phase reaction fresh reactant is continually transported to the electrode by diffusion from bulk solution.) The symmetry of the voltammetric peaks occurs since the electrode reaction is now controlled only by electron transfer kinetics (eqn 1.21) and not by a coupled problem of diffusion and electron transfer. This also explains why the maxima of the forward and reverse peaks now coincide. The areas of the two peaks are identical and provide a direct measure of the amount of charge required to perform the electrochemical reaction and, therefore, the quantity of material adsorbed on to the surface. An example of this type of behaviour is given by the adsorption of 2,5-dihydroxy-4-methylbenzylmercaptan (DMBM) on to a platinum (111) surface, Fig. 3.22. DMBM is known to strongly adsorb to electrode surfaces and has been used in spectroscopic studies (see Chapter 5). The observed voltammetry shows the ideal behaviour noted for a stable one-electron transfer reaction.

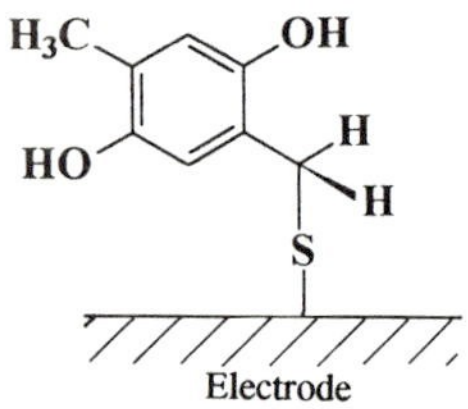

Deviation from ideal behaviour will occur if the surface-adsorbed species is not stable in either its charged or uncharged forms, or is desorbed for some reason during the potential cycling. Any deviation in the voltammetric behaviour can then be used to probe the stability and nature of the bond to the surface. For example, if, after initial electrochemical reaction, the species slowly desorbs, then a reduction in the reverse peak height would be observed. Variation of scan rate in an analoguous way to the homogeneous approach may then enable an estimate of the kinetics of desorption.

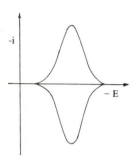

Fig. 3.21 Cyclic voltammogram of a surface-adsorbed species.

Fig. 3.22 Schematic of DMBM adsorbed on an electrode surface.

Data taken from Hubbard (1989).

Electrochemical studies of surface processes can also be extended to the investigation of adsorption on to single crystal faces. It has been demonstrated that, depending upon the site of adsorption, hydrogen bound to the surface can be oxidized at different potentials (Fig. 3.23). These ideas were also examined initially by surface techniques such as low energy electron diffraction (LEED) and Auger spectroscopy. There remains considerable debate on various aspects of these studies and interested readers should consult the recent review by Parsons and Ritzoulis (1991) for further details.

More recently scanning tunnelling microscopy (STM) has been employed to identify particular crystal planes accessible to adsorption on the surface.

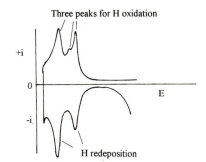

Fig. 3.23 Cyclic voltammogram of a polycrystalline Pt electrode in an aqueous solution of 0.5M H_2SO_4.

Chapter 5 discusses in more detail the basis and applications of STM, LEED, and Auger spectroscopy.

Modified electrodes

An electrode can be chemically modified, by the immobilization of a layer of material on to its surface. The modified electrode then takes on the surface character of the immobilized material and can be used to perform electrochemical experiments. Modified electrodes are found in a wide range of applications, since they offer novel properties such as the ability to hold charge (as in a battery), induce novel specific reactions, act as organic semiconductors and catalyse reactions.

Polymer-coated electrodes are just one example of modified electrodes and these are discussed as an example of some of the possibilities available from such an approach. Polymer-modified electrodes can be broadly divided into three main categories.

(a) *Electronically conducting polmers.* These polymers possess π states and are generally conjugated materials, they are often termed 'organic metals' (Fig. 3.24(a)).

(b) *Redox polymers.* The electrical conductivity occurs via self-exchange reactions within the polymer and with the electrons 'hopping' from one site to another (Fig. 3.24(b)).

(c) *Loaded ion exchange polymers.* These are dependent on both self exchange and physical diffusion of counter ions into the film (Fig. 3.24(c)).

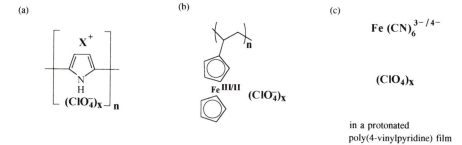

Fig. 3.24 Categories of electroactive polymers. (a) Polypyrrole, an electronically conducting polmyer; (b) poly(vinylferrocene), a redox polymer, and (c) iron ferro/ferri cyanide in poly(4-vinylpyridine), an example of a loaded ion-exchange polymer.

For a more detailed account of
polymer coated electrodes see
Murray (1984) or Bartlett (1994).

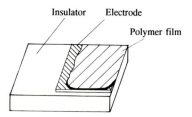

Fig. 3.25 Schematic representation
of a polymer-coated electrode.

Each of these categories possess a number of common features. All are
electroactive and thus can be either reduced or oxidized, depending on the
electrode potential. They have the ability to pass electrical current and are
generally mechanically semi-rigid. This last parameter is significant since the
electroactive polymer is usually a thin film deposited on the electrode surface.

Production of these modified electrodes can be performed by a variety of
techniques. In the case of electronically conductive polymers, the most common approaches are via casting from solution or direct electropolymerization.
The result is an electrode coated in a thin film of electroactive polymer (Fig.
3.25). Once chemically modified the electrode is capable of substantially
altered reactivity in comparison to the bare electrode. By way of example,
consider the reactivity of vitamin B_{12}, which is known to be an extremely
effective catalyst in biological systems. The modification of a carbon paste
electrode with a polymer incorporating vitamin B_{12} has been used to facilitate
the reductive, Michael addition of ethyl iodide to acrylonitrile (Fig. 3.26).

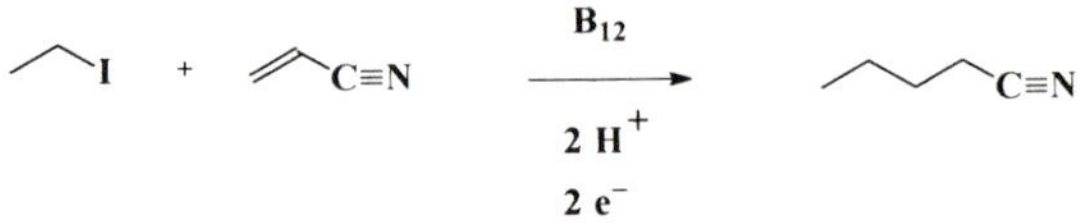

Fig. 3.26 The formation of valeronitrile using a polymer coated electrode incorporating
vitamin B_{12}.

The electrode coat contained approximately 1×10^{-10} mol cm^{-2} of vitamin
B_{12} and when potentiostatted at -1.4V (versus the saturated calomel electrode
SCE) the production of valeronitrile (A) occurred. The calculated catalytic
turnover of the modified electrode was found to be about 2100, demonstrating
the potential value of such devices.

Metallic deposition on electrode surfaces
Metallic cations exist in solution with a surrounding solvation shell and can
be adsorbed on to electrode surfaces, typically when the electrode surface
possesses a net negative excess charge. Depending upon the potential of the
electrode, cations can be reduced to lower oxidation states or discharged completely on the electrode, forming metal atoms on the surface in mono/sub-
mono layer quantities. This deposition can occur at potentials positive of that
predicted by the Nernst equation for bulk deposition processes. This effect is
termed underpotential deposition (upd) and arises primarily from a difference
between the chemical potential of the electroactive species in bulk solution
and that absorbed on the electrode surface. This occurs as a result of the
stronger interactions between the electrode/deposited metal than the
metal/bulk metal. The difference in the potential between bulk deposition and
surface deposition gives an indirect measurement of the binding energy of the
deposited metal to the electrode.

Of particular interest are the resulting electrical properties of the upd layer
(and therefore electrode), which differ significantly from those of the original

electrode material. The upd layer can also alter the nature of the potential distribution across the electrode/electrolyte interface.

Experimentally it is possible to monitor deposition effects using voltammetric means such as cyclic voltammetry, or by spectroscopic analysis (see Chapter 5). This gives direct information about the nature of the surface structure, both before and after the deposition process. Figure (3.27) shows an elegant example of the deposition of a metal on to the electrode surface. Considering, initially, only the forward sweep. As the potential is swept gradually, to more negative values a small, sharp peak is observed due to the deposition of a monolayer of lead on to the silver surface as a result of the upd processes. The larger feature at more negative potentials corresponds to the bulk deposition process and the electrode surface now possesses a bulk metal layer. The metal deposited at negative potentials can however be removed from the surface by sweeping to increasingly positive potentials. As the potential is increased the 'bulk' metal deposited on the surface is re-oxidized and dissolves back into solution. At a potential close to that observed on the reductive sweep a sharp features again observed corresponding to the desorption of the remaining monolayer of lead on the surface back into solution.

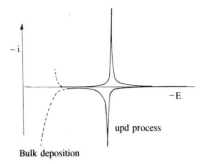

Fig. 3.27 Voltammogram of the deposition of lead on to a silver (111) surface.

Corrosion

Corrosion has been extensively investigated owing to the huge financial cost incurred by corrosion damage each year. In all cases corrosion occurs because of the instability of a material in relation to the environment around it. Corrosion generally occurs by the interaction of a substate surface with the environment, typically by a range of physical, chemical, or electrochemical interactions. The overall result of corrosion is the modification or degradation, of the surface. This degradation is often unpredictable and cannot necessarily be understood by simple thermodynamic properties. For example, the oxidation of aluminium to form Al_2O_3 is more thermodynamically favourable than the oxidation of iron to Fe_2O_3, yet aluminium is much more stable in general usage than iron (which is well known to rust). This can be rationalized, however, as aluminium forms a thin oxide layer at the surface of the metal which protects the surface from further oxidation.

The complexity of this particular area precludes a thorough review of the mechanisms that can occur. However to illustrate some of the ideas, the corrosion of metals in contact with water is considered.

Metallic corrosion generally occurs through oxidation of metal at the substate surface

$$M(m) \longrightarrow M^{n+}(aq) + ne^-(m)$$

Provided a high enough solution conductivity exists this reaction can be driven typically by a reaction, such as

$$2H_2O(aq) + 2e^-(m) \longrightarrow H_2(g) + 2OH^-(aq)$$

The oxidized metal can then either dissolve into the water or react with hydroxide at the surface to form an oxide or hydroxide layer across the

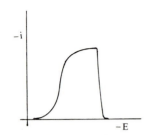

Fig. 3.28 Schematic of the voltammetric response of a metal undergoing corrosion.

surface. The formation of this surface layer can sometimes help to protect the metal below, making the material passive to further corrosion.

Electrochemical methods can be used to investigate corrosion. Perhaps the simplest approach is to monitor the metal

potential as the corrosion process occurs, at open circuit. The corrosion potential, $E_{corrosion}$ established can vary with time as the nature of the surface is altered by reaction.

Further information regarding corrosion processes can be obtained by recording the response of the metal to an applied potential. Figure 3.28 shows a schematic example of the voltammetric response of a metal that is capable of forming an oxide film which makes the metal passive to further corrosion. Initially as the potential is swept to positive potentials the metal begins to be oxidized (e.g. Fe to Fe^{2+}). The oxidized metal may then react with hydroxide in solution and start to form a film of oxide across the surface. At significantly oxidizing potentials the surface is completely covered by the oxide layer and the current drops abruptly as no further reaction can occur. The metal has then formed a coat which is non-conducting and therefore no current is observed in the voltammogram. At extreme positive potentials the film is finally destroyed by evolution of O_2 at the electrode, enabling current to flow again. The kinetics of the corrosion process may be obtained using Tafel analysis over the region of corrosion potentials (E_1 to E_2). Although such techniques can provide mechanistic and kinetic information about the corrosion process, some care must be exercised when analysing such data, as the current detected at a given potential may arise from a series of complex processes.

A further technique available for the investigation of corrosion processes is a.c. impedance spectroscopy. This technique can be used to provide information about film thickness, conductivity, and reactions at the interface. The principles behind a.c impedance and its use to investigate reactions are described in Chapter 4.

Photoelectrochemistry

Photoelectrochemistry can be split into two separate areas. First, light is absorbed into the electrode (typically a semiconductor) and this can induce changes in the electrode's conduction properties, so altering its electrochemical activity. Secondly, light is absorbed in solution by electroactive molecules or their reduced/oxidized products induce photochemical reaction and modification of the electrode reaction. In this section this second area is described, the photochemistry of semiconductor electrodes is detailed in Chapter 4.

Early experimental investigations employed an RDE apparatus with UV/visible transparent electrodes (e.g. tin oxide films). Photochemical excitation was achieved by shining the light source through the back of the transparent electrode. The photoelectrochemical reaction was monitored by comparison of the voltammetric response under 'dark' (non-irradiated) and irradiated conditions. Any difference detected in the current from the dark and light is called a photocurrent. The photocurrent can then be analysed using similar approaches to the investigation of electrode reactions as described in Section 3.3 (p. 33).

Using a similar procedure the photoelectrochemical mechanism for the expulsion of iodide from 1-iodoanthraquinone (Fig. 3.29) has been investigated. A combination of spectroscopic detection and photocurrent measurements were required to elucidate the reaction mechanism and consequently a channel electrode was used. Figure 3.30 shows the results for a typical current/voltage scan under steady-state mass transport conditions. The solid line shows the voltammetric behaviour of 1-iodoanthraquinone, which in the 'dark' undergoes a reversible one-electron reduction to the radical anion. When the electrode is irradiated with light at 560 nm, corresponding to the wavelength of an absorption band in the radical anion (but not the parent molecule), then an enhancement of the current is detected (shown by the dashed line). Chopping the light source on and off as the potential is swept to more reductive potentials the current is observed to jump rapidly between the solid and dashed lines, strongly indicating a photochemical rather than thermal reaction. Measurement of the photocurrent as a function of the mass transport rate enabled a unique mechanistic scheme for the reaction to be proposed (Fig. 3.31).

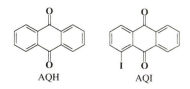

Fig. **3.29** anthraquinone (AQH) and 1-iodoanthraquinone (AQI).

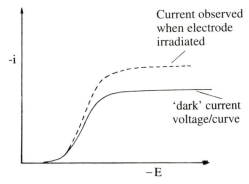

Current observed when electrode irradiated

'dark' current voltage/curve

Fig. 3.30 Steady-state voltammogram under chopped illumination.

$$AQI(aq) + e^-(m) \rightleftharpoons AQI^-(aq)$$

$$AQI^-(aq) \xrightarrow[\text{'H'}]{\text{irradiation}} AQH(aq) + I^-(aq)$$

$$AQI^-(aq) + AQH(aq) \rightleftharpoons AQI(aq) + AQH^-(aq)$$

$$AQH(aq) + e^-(m) \rightleftharpoons AQH^-(aq)$$

Fig. 3.31 The mechanism of reaction for the photoelectrochemical activation of 1-iodoanthraquinone.

Fig. 4.1 Schematic of (a) The Helmholtz electrical double layer model and (b) the potential drop across the interface.

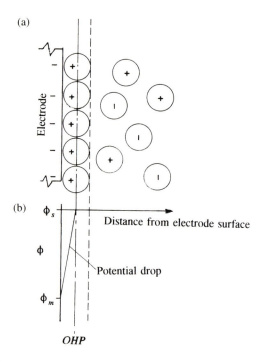

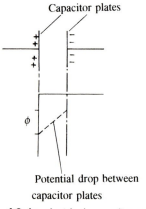

Capacitor plates

Potential drop between capacitor plates

Fig. 4.2 An electrical capacitor and the potential drop between the plates.

drawn through the centre of these ions at a minimum distance from the electrode surface is called the Outer Helmholtz Plane (OHP). In this simple model the excess charge on the metal is balanced completely in solution by ions situated at the OHP, and the potential drop across the interface occurs totally over the region between the metal surface and the OHP. During the esulting separate layers of charge, Helmholtz called the electrode/electrolyte interface the 'electrical double layer'. This double layer is equivalent to an electrical capacitor (Fig. 4.2) in which two layers of charge are separated by a fixed distance. The potential drop between these two charged layers is linear.

Helmholtz's model was improved by Gouy and Chapman who, working independently, concluded that the excess charge density in solution is not exclusively situated at the OHP. This occurs since the electrostatic attraction forces attracting or repelling ions from the electrode are counteracted by Brownian motion in solution which tends to disperse the excess ions. Their model used point charges to represent the ions and proposed that the charge density in solution is contained within a single 'diffuse layer' close to the electrode surface (Fig. 4.3). In this diffuse layer the net charge density decreases with distance away from the phase boundary. The potential drop across this diffuse layer is then mainly concentrated in the region closest to the electrode surface, but some charge is now held further away from the electrode than the OHP.

In 1924 Stern developed this approach further by assuming that the ions have a minimum distance of approach (the OHP) as well as accepting the idea of Gouy and Chapman that Brownian motionin solution created a diffuse

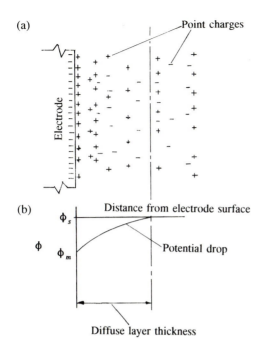

(a) Point charges

Electrode

(b)

ϕ_s

ϕ ϕ_m

Distance from electrode surface

Potential drop

Diffuse layer thickness

Fig. 4.3 Schematic showing (a) the Gouy–Chapman model of the electrical 'double layer' and (b) the potential drop across the electrode/solution interface.

layer. The result is effectively a combination of the two models. In terms of the potential distribution across the interface, a sharp drop exists between electrode (ϕ_m) and OHP (ϕ_{OHP}) beyond which the potential gradually falls to a value characteristic of the bulk electrolyte (Fig. 4.4).

In 1947 Grahame proposed that although the zone closest to the electrode is mainly occupied by solvent molecules it may be possible for some ionic or uncharged species to penetrate into this region. This could occur if the ion possessed no solvation shell, or if the solvation shell was lost when the ion approached the electrode. In this case ions are directly in contact with the electrode and are said to be 'specifically adsorbed'. The adsorption is termed 'specific' since the interaction occurs only for certain ions or molecules and is often unrelated to the charge on the ion. For example, it is possible for negatively charged ions to adsorb onto an electrode surface that already carries a negative charge density.

The model for the electrode/electrolyte interface was therefore modified and a new (closer) plane of minimum approach (Fig. 4.5) identified. This is termed the Inner Helmholtz Plane (IHP) and is defined as the axis through the centre of the specifically adsorbed species. The general effect of specific adsorption is to reduce the charge density needed in solution, in the 'double layer', to compensate the charge on the electrode.

More modern treatments of the double layer rely on the use of statistical mechanics and are outside of the scope of this text.

Although the more advanced models no longer consist of a simple double layer the term is still retained by the electrochemical community to refer to the electrode/electrolyte interfacial region.

Gouy (1910) suggested that specific adsorption may have some influence, but no account of it was made in his model.

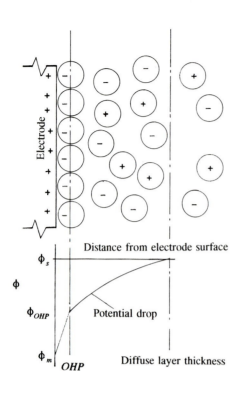

Fig. 4.4 The Stern model of the electrical double layer.

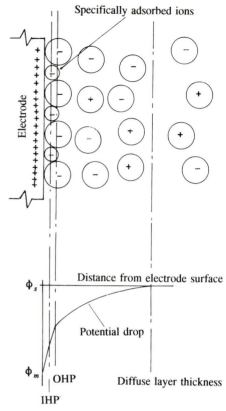

Fig. 4.5 Grahame model.

4.2   Probing the electrode/electrolyte interface

The interpretation of double layer effects requires the extension of chemical thermodynamics from the form familiar to the treatment of processes in one phase to that necesssary to facilitate the description of interfacial regions.

The aim of the following is to provide a quantitative mathematical description of the ion build-up within the interface region brought about by the electrode charge. This build-up is quantified by the 'surface excess', (Γ_B) of a species B defined as

$$\Gamma_B = \int_0^\infty ([B]_x - [B]_{bulk})\, dx = \frac{n_B - n_B^{bulk}}{A}$$

where A is the area of the interface region. $[B]_x$ is the concentration of species B at a point x in the double layer and $[B]_{bulk}$ the bulk concentration of B. Alternatively, n_B is the total number of moles of species B in the double layer and n_B^{bulk} is the number of moles of species B in a region of bulk solution having the same dimensions as the double layer. To fully describe the interface the surface excess of every component, B, within the interfacial region needs to be identified.

For simplicity the electrode is assumed to simply attract and repel ions in solution and that *no* electrolysis is assumed possible. Grahame called such an electrode an 'ideal polarized electrode' and found that the electrode that behaves closest to ideal behaviour in reality is the mercury electrode in aqueous solution. Much of the experimental data on the double layer have been obtained using mercury electrodes.

The first step in establishing the thermodynamics of interfaces is to consider the change in the internal energy (U) of the system when work (dw) is done on the interface. Application of the 1st and 2nd laws of a reversible thermodynamic change gives

$$dU = TdS + dw \qquad (4.1)$$

where T is the absolute temperature and S is the entropy of the system. It is instructive to identify the work done if the electrode potential, (E) is changed by an amount dE. The previous section revealed that the primary effect will be the attraction and repulsion of solution-phase ions to or from the electrode. Several effects contribute significantly to dw.

First the movement of a molecule B into the interface may increase the interfacial volume (V). The work done upon expansion of the interfacial volume is given by

$$dw = -PdV$$

where P is the pressure.

Secondly, the area of the interface, A may be altered by influx of molecules. If the area is altered by an amount dA, then the work involved to stretch the area is given by

$$dw = -\gamma dA$$

where γ is the interfacial tension.

The interfacial tension is simply the work done in stretching the interfacial region by unit area.

Thirdly, chemical work also occurs upon the introduction of a species B to the interface. The magnitude of this chemical work when dn_B molecules are introduced to the interface is given by

$$dw = -\mu_B dn_B$$

where μ_B is the chemical potential of the molecule B.

Finally, the application of a potential to the electrode alters the excess charge density (q^m) on the metal. The electrical work performed for a change in the excess charge density of dq^m is

$$dw = -A\Delta\phi_{m/s}dq^m$$

where $\Delta\phi_{m/s}$ is the *absolute* drop in potential at the electrode/solution interface induced by the change, dE, in the electrode potential.

Substituting all of these contributions to dw into eqn 4.1 gives the following result for the internal energy change

The chemical potential is the Gibbs free energy per mole, therefore $\mu.dn$ gives the free energy change. For a reversible system this is equal to the work done on the system (charge multiplied by potential).

$$dU = TdS - pdV - \gamma dA - A\Delta\phi_{m/s}dq^m - \sum_B \mu_B dn_B \tag{4.2}$$

Integrating gives,

$$U = TS - PV - \gamma A - A\Delta\phi_{m/s}q^m - \sum_B \mu_B n_B \tag{4.3}$$

The Gibbs free energy is defined as

$$G = U + PV - TS \tag{4.4}$$

Combining eqns 4.3 and 4.4 gives

$$G = -\gamma A - A\Delta\phi_{m/s}q^m - \sum_B \mu_B n_B \tag{4.5}$$

which on differentiation becomes

$$dG = -\gamma dA - Ad\gamma - A\Delta\phi_{m/s}dq^m - Aq^m d\Delta\phi_{m/s} - \sum_B (\mu_B dn_B + n_B d\mu_B) \tag{4.6}$$

Also, from eqn 4.4

$$dG = dU + PdV + VdP - TdS - SdT \tag{4.7}$$

and substituting eqn 4.2 gives

$$dG = VdP - SdT - \gamma dA - A\Delta\phi_{m/s}dq^m - \sum_B \mu_B dn_B \tag{4.8}$$

Comparison of eqns 4.6 and 4.8 shows that for constant T and P

$$Ad\gamma = Aq^m d\Delta\phi_{m/s} - \sum_B n_B d\mu_B \tag{4.9}$$

Equation 4.9 predicts variation of surface charge with the interfacial tension. It may be simplified into an experimentally helpful form known as the Gibbs–Lippmann equation

$$d\gamma = -dq^m dE - \sum_B \Gamma_B d\mu_B \tag{4.10}$$

In jumping from eqn 4.9 to eqn 4.10 the interfacial potential drop $\Delta\phi_{m/s}$ has been related to the electrode potential as discussed in Chapter 1, and the following identity employed

$$\sum_B \Gamma_B d\mu_B = \frac{1}{A}\left(\sum_B n_B d\mu_B - \sum_B n_B^{bulk} d\mu_B\right) = \frac{1}{A}\sum_B n_B d\mu_B \tag{4.11}$$

Equation 4.11 follows from eqn 4.9, since in bulk solution electric fields and surface forces are absent so that

$$\sum_B n_B^{bulk} d\mu_B = 0 \tag{4.12}$$

The Gibbs–Lippmann equation, shows that the charge density on the metal, and the surface excess in the double layer, control the interfacial tension. In particular, for the case of a solution of constant composition, $d\mu_B = 0$. It then follows from eqn 4.10 that

$$\left(\frac{d\gamma}{dE}\right)_{T,\rho,\mu_B} = -q^m \tag{4.13}$$

This is the important Lippmann equation, which relates the variation of the interfacial tension with the electrode potential to the charge on the metal. The former is experimentally measureable, as will be described in Section 4.3, so that the Lippmann equation gives a basis for the measurement of the charge density (q^m) on the metal and in the double layer since, $q^s = -q^m$.

Differentiation of eqn 4.13 leads to another very useful result

$$-\left(\frac{d^2\gamma}{dE^2}\right)_{T,\rho,\mu_B} = \left(\frac{dq^m}{dE}\right)_{T,\rho,\mu_B} = C_d \tag{4.14}$$

The quantity C_d, which shows how the charge on the electrode varies with the applied potential, is another quantity that is experimentally accessible as will be related in Section 4.3. As suggested by Section 4.1 C_d is the ratio of a charge to a potential and is known as the differential capacitance.

The thermodynamic analysis therefore provides two distinct possible experimental approaches for the investigation of the double layer. First, the measurement of the variation of surface tension with applied potential can be employed to provide information on the charge, q^s, held in the double layer. Secondly the variation of the differential capacitance with the applied potential provides an alternative method with which to study the double layer. Equation 4.14 shows that these two approaches are related and should give parallel results. In the following sections both these experimental methods are described.

4.3 Electrocapillary measurements

The variation of the interfacial surface tension as a function of the electrode potential provides a means of determining the charge in the double layer. Experimentally, 'electrocapillary' measurements are conveniently used for this purpose. These probe the mercury/aqueous solution interface.

Figure 4.6 shows a schematic of the basic instrumentation employed. A mercury-filled capillary is placed in contact with the electrolyte solution of interest and the mercury column. The potential of this electrode is controlled relative to a suitable reference electrode so that any alteration of the potential applied between the two electrodes results in a change in $\Delta\phi_{m/s}$ at the interface of interest. This two-electrode arrangement is satisfactory, since no electrolysis occurs at the water/mercury interface over a wide range of applied potentials.

The mercury-filled glass capillary is attached to a reservoir, the height of which may be adjusted above the mercury/electrolyte interface. The basis of

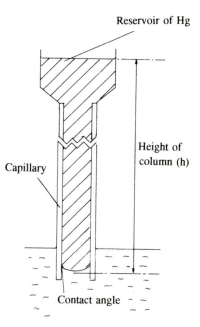

Fig. 4.6 Schematic of the mercury column employed for electrocapillary measurements.

the experiment lies in applying a potential to the mercury and varying the height of the reservoir until the mercury in the capillary is stationary and ceases to fall under gravity. Under these conditions the force of the interfacial tension is exactly balanced by the weight of the mercury column. The interfacial tension can then be calculated from a simple relationship

$$\gamma = \frac{h\rho g r}{2\cos\theta} \tag{4.15}$$

where h is the distance between the capillary and the reservoir, ρ the density, g the gravitational constant, and r the radius of the capillary (Fig. 4.6).

The above represents a simple approach to the measurement of interfacial tensions; more sophisticated approaches employ a controlled pressure to hold the position of the mercury interface. Nevertheless, the basic technique provides a sensitive probe of the interfacial region and has provided a wealth of data.

Using electrocapillary measurements, plots of interfacial surface tension against applied potential can be made. The gradient ($d\gamma/dE$) at any point on an 'electrocapillary curve' provides a measure of the surface charge by using eqn 4.13.

A typical electrocapillary curve for a mercury electrode in contact with an aqueous solution containing an electrolyte solution is shown in Fig. 4.7. The electrocapillary curve is approximately a convex parabola and the Lippmann equation (eqn 4.13) shows that the maximum of the curve corresponds to the point where the charge on the metal is zero. This has been (imaginatively) termed the 'potential of zero charge' (E_{pzc}).

As the electrode potential is changed from E_{pzc} the electrode becomes increasingly charged and ions in the electrolyte accumulate or are depleted at the interface so establishing a double layer. If $q^m > 0$ then anions will be attracted into the double layer and cations repelled, and vice versa if $q^m < 0$. The excess charge in the electrolyte layer is equal in magnitude, but of opposite sign, to the charge on the metal, so that overall electrical neutrality exists across the double layer. Experimental investigations using alkali and alkaline-earth fluorides have demonstrated this ideal behaviour. It was found that E_{pzc} was independent of the nature and concentration of the electrolyte (except at high concentrations). In many cases, however, E_{pzc} is much more sensitive to the electrolyte. Figure 4.8 shows electrocapillary curves for different electrolytes measured for concentrations of 0.1M.

The E_{pzc} values can be seen to vary considerably with the identity of the ions in solution. These experimental observations lead Grahame to propose the concept of the Inner Helmholtz Plane outlined in Section 4.1. This accepts that effects other than electrostatic forces or Brownian motion are present in the double layer and, in particular, specific adsorption can exist for both charged and uncharged species.

Close examination of Fig. 4.8 shows that the Data taken from Grahame (1947).electrocapillary curves are insensitive to the nature of the ions in solution when the electrode potential is more negative than E_{pzc}. In contrast, when the potential is more positive then the observed interfacial tension varies

The force due to gravity is equal to

$$mg = (\pi r^2 h\rho)g$$

The force due to the surface tension is given by

$$2\pi r\gamma\cos\theta$$

When the column of mercury is stationary then the two forces are in balance and eqn 4.15 is obtained. For the glass/mercury system a further simplification arises as the contact angle (θ) is approxiamtely equal to zero and thus eqn 4.15 reduces to

$$\gamma = \frac{h\rho g r}{2}$$

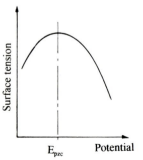

Fig. 4.7 Schematic of a typical electrocapillary curve.

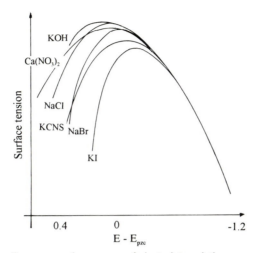

Fig. 4.8 Electrocaplliary curves for a range of electrolyte solutions.

markedly from one electrolyte to another. This contrasting behaviour has been explained in terms of the relative ease of dehydration of the cations and anions. In particular, apart from a few exceptions, such as F^- and SO_4^{2-}, anions are much more easily dehydrated than cations. Accordingly, at potentials more positive than E_{pzc} the interface contains dehydrated anions positioned at the Inner Helmholtz Plane. The size of this layer reflects the size of the (dehydrated) anion and consequently varies considerably from one anion to another. In contrast, cations are believed to adsorb with their hydration shells essentially preserved so that the ions are located at the Outer Helmholtz Plane. The size of the latter is therefore much less sensitive to the chemical identity of the cation and this explains why the electrocapillary curves in Fig. 4.8 appear to almost coalesce at negative potentials. These inferences are supported by the observation that E_{pzc} shifts in a positive direction for the series, I^-, Br^-, and Cl^-. This reflects the fact that for reasons of the charge density at the surface of the ion, I^- is much more easily dehydrated than Cl^-. As a consequence, cations are displaced from the electrode surface by iodide anions at less positive potentials than by chloride ions. Fluoride anions are thought to adsorb with their hydration shells intact and so E_{pzc} is most positive for electrolyte solutions containing F^-.

Equation 4.10 hints that it is possible to deduce surface excesses, Γ_B, via measurements of interfacial tension. The excess charge density for the particular ion can then be calculated

$$q^+ = z_+ F \Gamma_+ \qquad q^- = z_- F \Gamma_-$$

Figure 4.9 shows the variation of the surface excess as derived from Fig. 4.8 for several K^+ and I^+ ions as a function of the electrode potential.

Uncharged species can be found in excess within the double layer region as well as ions; many examples of the adsorption of organic molecules on to the electrode surface are known. These can interact with the electrode via

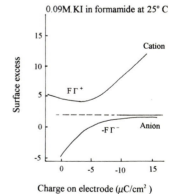

Fig. 4.9 Surface excess of cations/anions in the double layer.

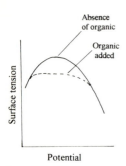

Fig. 4.10 The influence of organic adsorbants on electrocapillary measurements.

specific groupings within the molecule such as π-bonds. Alternatively, species that are less polar than the surrounding solvent may be encouraged to accumulate preferentially at the interface by their hydrophobic interaction with the solvent. In particular, tiny amounts of organic material can be found as trace levels of impurity in water and many of the early electrocapillary measurements were found subsequently to be in error due to adsorption of organic material on to the electrodes used. The adsorption is revealed by data such as that in Fig. 4.10 which shows that adsorption occurs preferentially close to E_{pzc}.

Surface tension has been shown to provide valuable information about the double layer. An alternative strategy relies on the measurement of the interfacial capacitance as a function of the potential.

4.4   Differential capacitance measurements

The models discussed above have shown that the double layer is a region in which charged electrolyte ions can be accumulated or depleted in comparison with their bulk values. This in effect, means that the interface is capable of storing charge and therefore acts like a capacitor. A simple parallel plate electrical capacitor is shown in Fig. 4.2. The integral capacitance (C) of the device is defined as

$$C = \frac{q}{E} \qquad (4.16)$$

where q is the quantity of charge 'stored' on the capacitor plates when a potential difference E volts is applied across them. For a simple capacitor the capacitance may be predicted using eqn 4.16 and is constant, independent of potential, since the separation of the plates is fixed.

The electrochemical cell represents a slightly more complicated problem as the capacity of the double layer is dependent upon the applied potential. This arises since, in effect, the thickness of the double layer changes with potential, the further the electrode potential is from E_{pzc} the greater the attraction or repulsion for the electrolyte ions and the smaller the dimensions of the double layer. In electrochemical measurements therefore the *differential* capacitance

$$C_d = \left(\frac{dq}{dE}\right) = \left(\frac{\partial q^m}{\partial \Delta\phi_{m/s}}\right)_{constant\ composition} \qquad (4.17)$$

is measured using a method called alternating current (a.c.) impedance spectroscopy. Equation 4.17 defines Cd first for a simple capacitor and, second, for the electrode/electrolyte interface. Cd varies with potential as the double layer changes. It will also change with the composition of the electrolyte; this explains the need for the subscript 'constant composition'.

4.5   A.C. impedance spectroscopy

A.C. impedance spectroscopy is widely used for the investigation of both solid and liquid-phase phenomena. The discussion that follows aims to provide a basic overview of the principles of the technique. First the response of a simple capacitor to an A.C. voltage is examined, which provides a model for the interpretation of real solution/electrode interfaces.

Experimental measurements are performed using an oscillating sinusodial potential, $E(t)$.

$$E(t) = E_m \sin(2\pi ft) \tag{4.18}$$

where f is the frequency measured in Hertz (s^{-1}) and E_m is the maximum amplitude. Consider Fig. 4.11 in which such a potential is applied to a simple capacitor. The oscillating potential across the cell causes a current to flow given by

$$i(t) = i_m \sin(2\pi ft + \theta) \tag{4.19}$$

where i_m is the maximum current amplitude and θ represents any phase difference between the applied voltage and detected current. The differential capacitance can be related to the current as follows

$$i\frac{dq}{dt} = C_d \frac{dE}{dt} = C_d\{2\pi f E_m \cos(2\pi ft)\} \tag{4.20}$$

This may be rewritten as

$$i = C_d\{2\pi f E_m \sin(2\pi ft + \pi/2)\} \tag{4.21}$$

from which it follows that the phase difference is $\theta = \pi/2$.

The response of electrical circuits to a.c. voltages is often described in terms of the impedance (Z) defined as

$$Z(f) = \frac{E(t)}{i(t)} \tag{4.22}$$

For the case of the capacitor under consideration, $Z = 1/(2\pi f C_d)$, and it may be appreciated that Z provides a measure of how the passage of current is 'impeded' by the circuit. As such the impedance provides a generalization of the idea of the 'resistance' of the circuit.

In the case of a capacitor the phase difference between current and voltage is $\pi/2$, (Fig. 4.12) whereas for a single resistance there is no phase difference and $\theta = 0$. In simple terms, this arises since charge can flow directly through a resistor, but with a capacitor the conduction is indirect and results from the attraction or repulsion of electrons from the plates as their potential changes. In any real, complex circuit the phase difference between current and voltage can take on any value and, as hinted by eqn 4.19, this can change with frequency. Mathematically

$$Z(f) = Z' \sin(2\pi ft) - Z'' \cos(2\pi ft) \tag{4.23}$$

Impedance measurements can therefore be conveniently described by a plot such as Fig. 4.13 in which the quantities Z' and $-Z''$ appear on the x- and y-axes, respectively. Typically, the impedance, Z, at each frequency is plotted

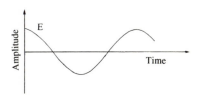

Fig. 4.11 The form of the applied potential used in a.c. impedance spectroscopy.

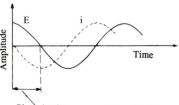

Phase lag between current and voltage

Fig. 4.12 The current response with a phase difference of $\pi/2$.

Electrical circuit

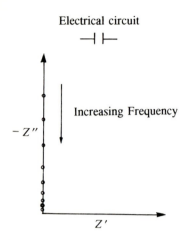

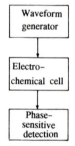

Fig. 4.13 Impedance plot for an electrical circuit containing a single electrolytic capacitor.

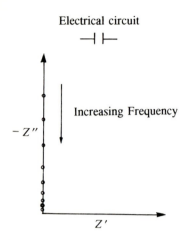

Fig. 4.14 Schematic of a typical apparatus employed for impedance measurements.

as a point with coordinates $(Z', -Z'')$. The resulting vector from the origin to this point represents the impedance. It has a magnitude equal to the length of this vector and the phase difference between the current and voltage is given by θ.

The shape of impedance plots such as that shown in Fig. 4.13 provide a very sensitive diagnostic of the circuit under investigation. It will be seen later that they are very useful in characterizing electrochemical systems in general. For the purposes of studying the electrode/solution interface it is helpful to consider the impedance plot for a simple capacitor. More complex situations are examined in Section 4.7.

Figure 4.13 shows the impedance plot for a simple capacitor. From eqn 4.23 the following expressions for the 'in-phase' (Z') and 'out-of-phase' $(-Z'')$ components may be written

$$Z' = 0 \qquad -Z'' = \frac{1}{(2\pi f C_d)}$$

It follows that the impedance plot takes the form of a vertical line along the *y*-axis $(Z'=0)$. Low frequency plots lie towards the top of the line and high frequency points towards the bottom.

Modern experimental techniques employ an electrochemical cell connected to a frequency response analyser (Fig. 4.14) which can generate impedance plots directly. The use of a three-electrode system and suitable reference electrode permits the impedance of a specific electrode/solution interface to be probed. The apparatus employs an electronic device to generate frequencies in the range 10^{-2} to 10^5 Hz. The ability to vary the frequency so widely enables a huge range of time-scales to be probed, although the early experiments in this area used the fixed operating frequency dictated by experimental contraints to probe the double layer region.

4.6   Differential capacitance measurements

The previous section revealed a.c. impedance Spectroscopy as a sensitive method of interrogating the electrode/solution interface. In particular, the apparatus shown in Fig. 4.14 permits the measurement of the differential capacitance of the interface

$$C_d = \left(\frac{\partial q^m}{\partial \Delta \phi_{m/s}} \right)_{\text{constant composition}} \tag{4.24}$$

If the electrode potential is insufficient to induce any electrolysis then no d.c. (or 'Faradaic') current will pass. However, a fluctuating a.c. voltage will induce ion redistribution within the double layer. As the voltage changes in a positive direction, anions will be attracted, and cations repelled, from the double layer. Then, as the voltage swings in the opposite sense, the converse takes place. In this manner an a.c. current is induced to pass through the interface although no electrons actually cross the electrode/solution boundary. This is termed a charging (or 'non-Faradaic') current. As the latter is a direct result of the redistribution of ions within the double layer the experimental technique provides an insight into the processes occuring in the double layer.

Following the simple approach of Stern two capacitances can be associated with the double layer. The first, C_{OHP}, is the differential capacitance of the OHP and the second, $C_{Diffuse}$, corresponds to that of the diffuse region. The overall capacitance is then simply given by the sum of the two

$$\frac{1}{C_d} = \frac{1}{C_{OHP}} + \frac{1}{C_{Diffuse}} \qquad (4.25)$$

When no specific adsorption occurs then C_{OHP} is independent of the electrolyte concentration as only solvent molecules are present within this layer. In contrast $C_{Diffuse}$ is very sensitive to the ion excesses or depletions within the diffuse layer. Depending upon the electrolyte concentration and applied potential, the capacitative properties of the double layer will be effected in different proportions by C_{OHP} or $C_{Diffuse}$. Note that since the total capacitance depends on the *reciprocals* of the two separate contributions it is the smallest capacitance that dominates C_d.

Figure 4.15 shows the behaviour of the differential capacity of an aqueous solution containing varying concentrations of NaF. The basic features of this plot can be understood by considering the effects of the different concentrations of electrolyte. At low concentrations of electrolyte (0.001M in Fig. 4.15) and potentials close to E_{pzc}, the thickness of the diffuse layer is large, as only a weak attraction from the electrode exists and low concentrations of the electrolyte exist in solution. The measured differential capacitance C_d is therefore small

$$\frac{1}{C_{OHP}} \ll \frac{1}{C_{Diffuse}} \qquad (4.26)$$

and from eqn 4.25 $C_d \approx C_{Diffuse}$. As the applied potential is gradually shifted from E_{pzc} the electrostatic attractive force becomes greater and this acts to compress the diffuse layer closer into the electrode. This gives rise to an increase in the capacitance. Eventually $C_{diffuse}$ becomes large in comparison with C_{OHP} so that

$$\frac{1}{C_{OHP}} \gg \frac{1}{C_{Diffuse}} \qquad (4.27)$$

and $C_d \approx C_{OHP}$. This explains the behaviour at extreme potentials positive and negative of E_{pzc} since under these conditions C_d will be approximately given by $C_d = \varepsilon_o \varepsilon_r A/d$, where ε_o is the relative permittivity free space and ε_r the relative permittivity of the medium of interest Consequently, will be essentially independent of potential but reflect the thickness, d, of the Helmholtz Layer. At the positive potentials shown in Fig. 4.15 fluoride ions will dictate this thickness, whereas at negative potentials sodium ions will line up on the OHP. This explains why C_d approaches different limiting values at the two extremes.

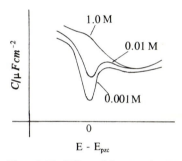

Fig. 4.15 Differential capacitance plots for varying concentrations of NaF.

Data taken from Grahame (1947).

A full discussion of the form of the expression

$$C_d = \frac{\varepsilon_o \varepsilon_r A}{d}$$

may be found in Atkins (1994).

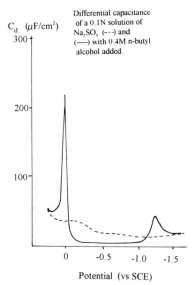

C_d ($\mu F/cm^2$)

300

Differential capacitance of a 0.1N solution of Na$_2$SO$_4$ (- - -) and (——) with 0.4M n-butyl alcohol added.

200

100

0   -0.5   -1.0   -1.5

Potential (vs SCE)

Fig. 4.16 The influence of organics on the differential capacitance.

Electrical circuit

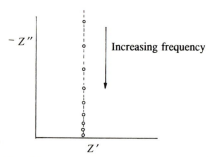

$-Z''$

Increasing frequency

Z'

Fig. 4.17 Impedance plot for resistor and capacitor in series.

At high concentrations of electrolyte the minimum seen in the low concentration case is not observed, as enough electrolyte is always present close to the electrode to ensure that the diffuse layer contribution is negligible and C_d C_{OHP}. Note though that at potentials positive of E_{pzc} the value of C_d again reflects the size of the Helmholtz Layer defined by F$^-$ ions at the OHP, whereas at negative potentials the latter are replaced by Na$^+$ ions.

In addition to the effects of electrolyte in the absence of any specific adsorption, a monomolecular layer of water exists at the electrode surface. Water molecules possess large dipole moments which can respond to electric fields. When a large charge density is present on the electrode (q^m) a very large electric field is created over the Helmholtz Layer which can fix these water molecules into a 'rigid' orientation. Differential capacity measurements can also provide information about specific adsorption. Figure 4.16 shows the influence of an adsorbed organic molecule, *n*-butyl alcohol, on the differential capacitance of an aqueous solution containing 0.1M Na$_2$SO$_4$. For comparison, the dotted line shows the electrolyte-only case. At potentials close to E_{pzc} the adsorbed organic lowers the differential capacitance from that seen for sodium sulphate alone. The adsorbed organic layer is uncharged so that $C_{OHP} = 0$ and this is reflected in the observed experimental behaviour.

As the potential is moved away from E_{pzc} the organic material is observed to desorb over a small potential range. This is observed at potentials both positive and negative of E_{pzc}. The desorption is revealed by very sharp spikes in the potential/capacitance plot. The behaviour after these peaks returns to that of the normal electrolyte solution as the organic molecules are lost into bulk solution.

The above discussion shows how impedance measurements made as a function of frequency can reveal the capacitative nature of the electrode/electrolyte interface. The latter is revealed by a vertical impedance plot as discussed in Section 4.5 and analysis of the frequency dependence of Z'' permits the inference of values of C_d. This illustrates a general approach to the analysis of impedance data; an equivalent circuit made up of resistors and/or capacitors is tested as a model against experiment and the results used to provide insights into the behaviour of the system under study.

4.7   Equivalent circuits for the interpretation of Electrochemical cells

It is informative to consider the impedance response of some simple electrical circuits, noting that in Section 4.5 the impedance plots for a simple capacitor and resistor were described. In particular, the behaviour of systems containing more than one simple component is examined. Figure 4.17 shows the impedance plot observed from a resistor (R ohms) and capacitor (C Farads) in series. A line is seen corresponding to

$$Z' = R \qquad -Z'' = \frac{1}{2\pi f C_d} \qquad (4.28)$$

These coordinates plotted for different frequencies, generate a line parallel to the Z'' axis and offset from it by an amount R/Ω. It is evident that the plot is simply the sum of those previously inferred for the two components measured separately. Analysis of the plot permits the direct measurement of R and the inference of C by plotting $(-Z'')$ against $(1/f)$.

The impedance behaviour becomes slightly more complicated when the resistor and capacitor are placed in parallel. In this arrangement the current can flow through either or both of the components, whereas in that case it is compelled to pass through both. In particular, it was noted that the ease of current flow through the capacitor is greater at higher frequencies (eqn 4.21). Accordingly, at low a.c. frequencies the current passes preferentially through the resistor, whereas at higher frequencies it flows predominantly through the capacitor which displays a negligible impedance. The total impedance thus approaches a value R on the x-axis at low frequencies but decreases to zero as f gets larger. The impedance plot is a characteristic semi-circle (Fig. 4.18).

The third case of interest has a resistor (R_2) in series with the parallel RC combination discussed above. In this case the current is obliged to pass through R_2 and then splits itself between the remaining two components. It follows that even at high frequencies the current will be subject to an impedance of R_2 ohms so that the impedance plot again takes the form of a semicircle as in Fig. 4.19 but this is now shifted by R_2 along the x-axis.

It is now possible to turn to a consideration of equivalent circuit models for an electrochemical cell. Consider a working electrode in a conventional three-electrode cell and assume that no electrolysis occurs. The simplest electrical representation of the system comprises a capacitor, corresponding to the double layer at the electrode, in series with a resistor, to mimic the solution resistance as shown in Fig. 4.20. This is exactly equivalent to Fig. 4.17 and the impedance plot consists simply of a vertical line offset from the Z''-axis. The values of R and C are readily inferred from the frequency dependence of the impedance.

Figure 4.21 shows a typical impedance plot obtained underconditions when an electrolytic reaction is taking place at the electrode. This may be understood using the Randles circuit shown in Fig. 4.22. In comparison with the no-electrolysis case, a resistance, R_{ct}, in parallel to the double layer capacity, C_{dl}, has been added to model the faradaic charge transfer reaction. A second, new term, the Warburg Impedance, allows for the frequency dependence of diffusive transport to the electrode. At high frequency the impedance plot shows that the dominant contribution to the total impedance is simply that of the solution resistance, since the double layer provides a path of negligible resistance to the current. It follows that at high frequency no electrolysis takes place. However, as the frequency is reduced this is no longer the case and the effect of R_{ct} in parallel with the C_{dl} gives rise to the characteristic semicircular part of the plot. Finally, as lower frequencies are reached the impedance shows a large rise, modelled by the Warburg impedance. This is due to significant concentration changes induced by the a.c. current which become increasingly difficult to replenish by diffusion as f decreases.

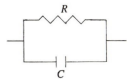
Electrical circuit

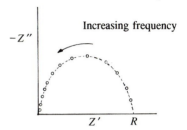

Fig. 4.18 Impedance plot for a resistor and capacitor in parallel.

Electrical circuit

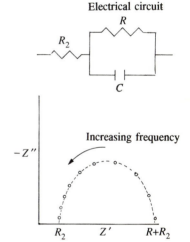

Fig. 4.19 Impedance response for a parallel RC circuit in series with a resistor.

Electrical circuit

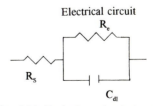

Fig. 4.20 Equivalent circuit for the influence of solution resistance on the impedance spectrum.

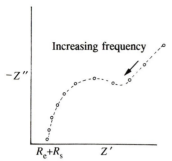

Fig. 4.21 Impedance plot showing the addition of an electrolytic reaction.

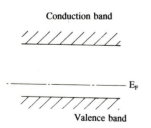

Warburg impedance

Fig. 4.22 The Randles circuit.

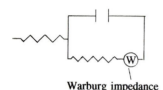

Fig. 4.23 Schematic of an intrinsic semiconductor.

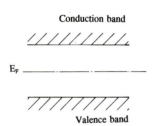

Fig. 4.24 Schematic of a p-type semiconductor.

The general approach discussed above has also been employed to study coupled homogeneous reactions and surface adsorption processes. A wide range of processes can also be studied such as corrosion, characterization of grain boundaries in solids, and general structural properties of materials. Impedance spectroscopy has also been applied to the study of the properties of semiconductor electrodes.

4.8 Semiconductor electrodes

The previous discussion has been concerned exclusively with metallic electrodes. Nevertheless, semiconductors may also be employed as electrodes and show some significantly different properties to their metallic counterparts.

Semiconductor materials, such as silicon, possess energy levels in the form of valence and conduction bands, which are separated by an energy gap. Conduction requires that some electrons in the valence band are excited into the conduction band either by thermal or photochemical excitation. Upon excitation an unoccupied vacancy (a hole) is left in the valence band. The holes and excited electrons can move in response to an applied electric field and so permit the passage of current.

A schematic diagram of an intrinsic semiconductor is shown in Fig. 4.23. In such semiconductors the holes and electrons exist in the pure material and semiconduction occurs without the introduction of any other material.

Semiconduction can be controlled through the deliberate introduction (doping) of small quantities of material to an intrinsic semiconductor, so forming an extrinsic semiconductor. The additive can be either electron donating or electron accepting. Suppose that a dopant atom with a filled energy level close to that of the conduction band is introduced into an intrinsic semiconductor. Donation of negatively charged electrons to the conduction band will be enhanced and an n-type semiconductor results, which has an enhanced conductivity over the corresponding intrinsic material. Conversely, if an electron-accepting dopant is added with an empty energy level just above the valence band then electron donation from the valence band to the empty level will take place resulting in a greater population of positively charged holes in the former and the formation of a p-type semiconductor (Fig 4.24).

When a semiconductor solid is brought into contact with a solution containing a redox couple then electron transfer may take place between the solid and the solution-phase species until the electrochemical potential of the two phases becomes equal. When equilibrium is established the semiconductor will have gained a net positive or negative charge. This charge resides in a region near the surface of the solid and causes an electric field to be established within the semiconductor. This region is known as the space–charge layer and has a typical thickness of between 2 and 500 nm, depending on the conductivity of the solid.

In order to describe the distribution of charge within the semiconductor the semiconductor bands are thought of as 'bent'. This implies an excess of negative charge at the surface if the bands are bent downwards, or an excess of positive holes at the surface if they bend upwards.

Figure 4.25 shows the band-bending effect induced by a redox couple (undergoing charge transfer) for an intrinsic semiconductor. If the semiconductor is controlled by a potentiostat then the applied potential from the latter can change the energy of the conduction and valence bands in the bulk semiconductor. As a result the extent of the band bending within the solid can be changed and charge carriers brought to, or removed from, the space–charge layer so permiting electrolysis to occur at the solid/liquid interface. Consider the case of an n-type material. The application of a negative potential will encourage the build-up ('accumulation') of charge carriers at the electrode surface and reduction of solution species may occur. In contrast the application of a positive potential will remove ('deplete') charge carriers and oxidative processes are enhanced, relatively. For p-type material, negative potentials will deplete the space–charge layer of charge carriers (holes) whereas positive potentials will lead to their accumulation.

It was noted above that electron excitation into the conduction band of a semiconductor can be achieved by photoexcitation with the result that a hole is left in the valeance band. It follows that the electrochemical behaviour of semiconductor electrodes will be strongly modified if they are irradiated with light of an energy equal to, or greater than, that of the band gap. This is illus-

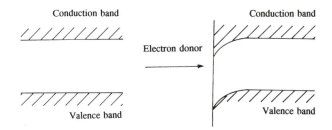

Fig. 4.25 The band-bending effect in semiconductors.

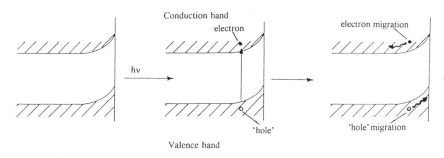

Fig. 4.26 Photoelectrochemistry.

trated in Fig 4.26 for the case of TiO_2. For an extrinsic n-type semiconductor, such as TiO_2, the conduction and valence bands are bent upwards at the surface unless a very negative potential is applied to the electrode. Consequently, those electrons photochemically promoted to the conduction band will be swept into the bulk of the material by the electric field present in the space–charge layer. Equally, the holes left in the valence band will move under the influence of the same field but will migrate to the surface of the solid. As a result, under conditions of irradiation, photooxidations are encouraged to occur at the interface of the n-type semiconductor and the solution. Examples for the case of n-TiO_2 include the photo-Kolbe reaction (Fig. 4.27).

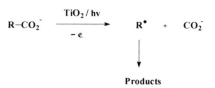

Fig 4.27 The photo-Kolbe reaction.

Bibliography

Atkins P.W. (1994), *Physical Chemistry,* OUP.

Bard A.J. and Faulkner L.R. (1980), *Electrochemical Methods, Fundamentals and Applications,* Wiley, New York.

Bockris J. O'M. and Reddy A.N. (1970), *Modern Electrochemistry,* Plenum, New York.

Chapman D.L. (1913), *Phil. Mag.,* **25**, 475.

Gouy G. (1910), *Compt. Rend.,* **149**, 654.

Grahame D.C. (1947), *Chem. Rev.,* **41**, 441.

Grahame D.C. (1947), *Ann. Rev. Phys. Chem.,* **6**, 337.

Helmholtz H.L.F. von (1853), *An n. Physik,* **89**, 211.

Stern O. (1924), *Z. Electrochem.,* **30**, 508.

5 Spectroelectrochemistry

The procedures described in Chapters 2–4 show how a mechanistic picture of an electrode reaction may be deduced using voltammetric information. However, the molecular identity of the species involved, particularly intermediates, must often be indirectly inferred. To provide more direct characterization electrochemical studies are frequently employed in combination with spectroscopic measurements. The additional information obtained from spectroelectrochemical measurements open the door to the investigation of a wide range of complex surface and homogeneous processes occurring in electrochemical systems.

Combined spectroscopic and electrochemical studies have benefitted greatly from advances in instrumentation over the past three decades and examples of spectroelectrochemical techniques can now be found utilizing radiation throughout the electromagnetic spectrum (Fig. 5.1).

The application of spectroelectrochemical techniques can be broadly divided into three areas

(a)   The structure characterization of the electrode surface.
(b)   The identification of homogeneous phase molecules.
(c)   The study of species adsorbed at the electrode/electrolyte interface.

As will become evident, many of the individual spectroelectrochemical techniques described below find application in more than one of the above areas.

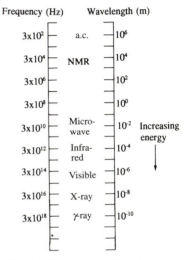

Fig. 5.1 Electromagnetic spectrum

5.1 Surface characterization using low-energy electron diffraction, Auger, and X-ray photoelectron spectroscopies

The ideal characterization of electrode surfaces requires information at, or approaching, atomic resolution. This can be achieved by the interaction of light or electron beams with the sample surface; the wavelength of the probe usually defines the surface resolution that can be achieved.

Low-energy electron diffraction (LEED), Auger, and X-ray photoelectron spectroscopy (XPS) are well-established procedures for the analysis of solid surfaces *in vacuo*. The three techniques are surface specific since each relies on the detection of electrons leaving the solid. The distance over which electrons may escape from a solid surface is of the order of 0.1–1.0 nm and therefore electrons detected due to interaction with the surface provide information about the first few layers of the sample surface. The three techniques are

Processes induced by x-ray excitation

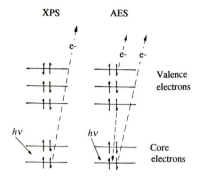

essentially complementary in the information acquired and therefore a combination of the three is optimally employed for the investigation of surface properties.

LEED probes the long-range structural order of the surface through diffraction of electron beams incident on it, analogous to that of X-rays by bulk crystals. The method is particularly valuable in characterizing electrodes made of single crystals. Typically, as with the other two techniques, the surface is examined *in vacuo* before and after electrolysis. The presence of adsorbates may also be probed provided a sufficiently high coverage is present and that specific sites are occupied so as to create long-range order.

In XPS X-rays focused on the electrode surface induce the direct ionization of surface atoms and the energy of the emitted photoelectron fingerprints the atom from which it was emitted and provides information about the oxidation state of the latter.

Auger electrons may also be emitted as a result of X-rays impinging on a surface; the loss of a core electron may be followed by electronic rearrangment within the surface atom and the secondary loss of a less strongly bound 'Auger' electron. The energy characteristics of the emitted Auger electron indicate the atomic composition of the surface.

All three experiments share a common requirement for performance in ultra high vacuum (UHV, pressures of approximately 10^{-10} Torr) to preclude surface contamination by the atmosphere. Accordingly, considerable effort has been directed towards the development of techniques and instrumentation that allow the surface preparation, electrochemical experiment, and surface characterization to be performed under ultra clean conditions. Figure 5.2 shows a general schematic diagram of the apparatus used in such electrochemical surface studies which employs several clean chambers interconnected via vacuum seals.

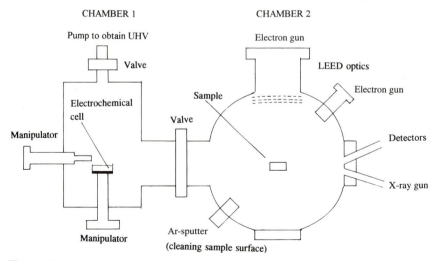

Fig. 5.2 Schematic of the general apparatus employed for vacuum surface studies.

Typically, a single crystal electrode is placed into chamber 1 where it is cleaned by bombardment from a beam of argon ions directed at the surface. The freshly prepared electrode is analysed using the techniques discussed above to provide information about the surface lattice structure and the levels of any impurities present at the surface.

The characterized electrode is then removed from the detection chamber to chamber 2 where the electrochemical experiment can be performed. A three-electrode cell is generally used under an inert atmosphere, such as ultra pure argon. The solvent is carefully purified before use and forms a protective barrier between any remaining levels of impurities present in the chamber and the electrode surface. Electrochemical experiments can then be performed on a well-characterized crystal surface. This may involve simple potential sweeps, which monitor the voltammetric response of the electrode or more complex processes such as upd or adsorption of molecules from bulk solution.

After electrochemical experiments the electrode is removed from the electrochemical cell maintaining potentiostatic control and returned to chamber 1, where further surface measurements can be performed. These probe the effects of the electrochemical experiment on the surface structure and, in addition, reveal the identity of any material that may have been adsorbed on to the electrode surface during the electrochemical experiment.

An enormous variety of surface processes have been studied using the above approach. Two examples are discussed below.

In one such study (Hubbard 1989) the electrodeposition of lead from an aqueous trifluoroacetic acid solution on to a well-defined and characterized silver (111) surface was investigated. Cyclic voltammetry measurements were performed on a single crystal Ag(111) electrode in solutions containing approximately 10^{-4}M concentration of lead acetate and 10^{-2}M trifluoroacetic acid. A typical cyclic voltammogram obtained is shown in Fig. (5.3). On the reductive scan three lead upd peaks (1,2,3) were observed at -0.34, -0.36 and -0.38 V, followed, at potentials more negative than -0.54 V by the bulk deposition (4) of lead on to the surface. Reversal of the sweep at the point of the bulk deposition gave rise to a peak corresponding to bulk lead dissolution (4'), followed by three further voltammetric features corresponding to the dissolution of the UPD processes (1',2',3'). Analysis of the charge contained in the peaks 1–3 allowed the packing density of lead to be calculated as $\theta = 0.58$.

After voltammetric study the electrode was removed from the electrolyte under potentiostatic control and transferred to chamber 1, where XPS, Auger electron, and LEED spectroscopic measurements were recorded.

The latter two techniques were used to confirm the voltammetric inference of lead deposition. An Auger electron spectrum (AES) is shown in Fig. 5.4 and reveals peaks due to silver and, also, to smaller quantities of lead at the surface. In addition, quantities of oxygen were found to be present at the electrode, as noted in Fig. 5.4). Analysis showed the surface packing density of oxygen to be approximately $\theta = 0.54$. The AES signal indicated that the spontaneous conversion of lead to lead oxide takes place on the surface when the electrode is removed at open circuit from solution.

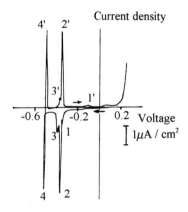

Fig. 5.3 Cyclic voltammogram of a single Ag (111) electrode in an aqueous solution of lead acetate and trifluoroacetic acid.

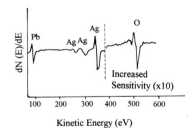

Fig. 5.4 AES spectrum of a Ag (111) electrode after the electrodeposition of lead.

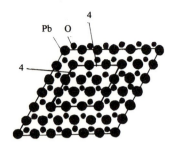

Fig. 5.5 Structure of lead oxide on a Ag (111) surface.

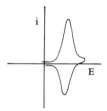

Fig. 5.6 Cyclic voltammetry of DMBM adsorbed on the electrode surface.

On the basis of the AES measurements and from further electrochemical studies, the product of the open circuit oxidation reaction was proposed to be lead oxide.

$$Pb + H_2O \rightleftharpoons PbO + H_2$$

Finally, LEED measurements were performed after the sample had been potentiostated just beyond −0.38 V. These revealed a (4 × 4) pattern and the structure of the lead oxide surface was consistent with that shown in Fig. 5.5.

The combination of voltammetry with surface science has also been used to probe organic molecules chemisorbed at an electrode surface. A good example is that of 2,5-dihydroxy-4-methylbenzylmercaptan (DMBM) anchored to a platinum (111) surface (Hubbard 1988). DMBM was shown to undergo reversible oxidation/reduction in the adsorbed state; Fig. 5.6 shows a cyclic voltammogram of DMBM in an acetonitrile solution. The forward and reverse peaks are symmetrical and the peak current occurs at near identical potentials for the oxidative and reductive processes.

Preliminary surface structural inferences were made using the voltammetric data as follows. First, measurement of the charge passed in the cyclic voltammogram enabled the coverage, in terms of moles per unit area, to be calculated. Secondly, the size of DMBM was calculated using a molecular unit cell approach in which the covalent and van der Waals radii of the constituent atoms (Fig. 5.7(a)) was used to estimate the area of each of the three faces (a, b and c) of the approximately box-shaped adsorbate. Finally by assuming that DMBM was attached as a monolayer over the whole surface of the electrode, a comparision of the observed coverage with that calculated for DMBM using each of the estimated facial areas showed that DMBM was attached to the electrode via face b.

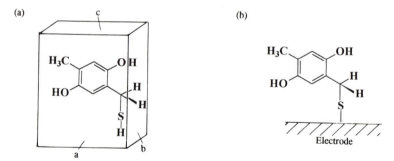

Fig. 5.7 (a) The unit cell of the DMBM molecule used in the calculations and (b) the proposed bonding interaction with the electrode surface.

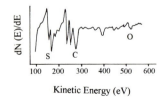

Fig. 5.8 Auger data from DMBM.

After voltammetric examination the electrode was transferred into the UHV chamber for surface characterization. Auger spectroscopy confirmed the identity of the attached material and the intensity of the Auger signal used (Fig. 5.8) to provide an independent estimate of the coverage of DMBM. The Auger data were found to be consistent with the voltammetric measurements and confirmed DMBM as bonded to the surface via face b.

The wealth of information available to the electrochemist from surface science techniques will be apparent. However, it will be recognized that all these techniques are of necessity performed outside the electrochemical cell. The *ex situ* aspect imposes limitations on the general applicability of these techniques and raises questions about the extent to which the process of transfer from cell to vacuum may influence the observations made. As a result experimental procedures have been developed to permit the *in situ* study of electrode processes. In the following sections some such techniques are illustrated.

5.2 In-situ atomic scale imaging of the electrode/ electrolyte interface using scanning tunnelling microscopy.

The basic principles of the scanning tunnelling microscope (STM) were identified in 1983 by Binnig and Rohrer. The major feature is the ability to image conducting surfaces at atomic resolution using a sharp 'tip' which is scanned across the substrate (Fig. 5.9) using a piezoelectric translator which defines the distance between the probe and the surface. A voltage is applied to the tip so that a tunnelling current flows between it and the surface. This tunnelling current is *extremely* sensitive to the separation of the two conductors and it is this sensitivity that provides the key to the resolving power of the experiment. In particular the STM can respond to changes of 0.01Å in vertical separation between tip and surface, so that individual atoms are visible in the latter as it is scanned.

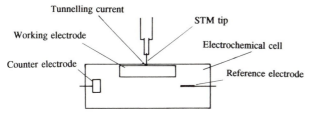

Fig. 5.9 Schematic of STM apparatus.

Typically, in experimental practice, as the probe is scanned laterally across the sample the tunnelling current is maintained at a constant value by the use of a carefully designed electronic feedback loop. The frequency of response of the feedback loop greatly exceeds that of any vibrations present in the surroundings so that the latter do not interfere with the experiment. As the probe encounters a change in surface topography the tunnelling current alters and the feedback loop then acts to re-establish the original tunnelling current. This is achieved by altering the voltage between tip and sample. This change in voltage is recorded and used to create an image of the surface (Fig. 5.10). The probe is fully scanned across the surface in the *x*-direction and then moved a small distance in the *y*-direction before a further scan is performed. The result

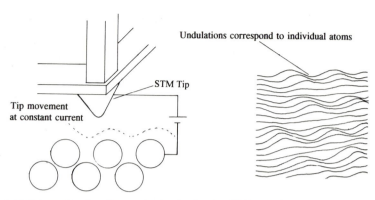

Fig. 5.10 Schematic of an image obtained using the STM technique.

is a pattern that represents the surface topography of the sample at atomic scale resolution. The absolute resolution in the z-direction is 0.01Å. The lateral resolution is dependent on the tip diameter. To obtain atomic resolution in this direction the current flowing to or from the tip must come from only one atom. It is often found that at the tip of the probe the required single atom exists and atomic resolution is achievable.

Initial STM experiments were performed at gas/solid interfaces. However, it soon became apparent that the technique could equally well be employed in fluid media without a significant loss in resolution. In the late 1980s, STM was applied to electrochemical investigations and a schematic diagram of a typical cell used is shown in Fig. 5.9. The probe tip and solid substrate are subject to individual potentiostatic control but the experiment otherwise proceeds as outlined above. The electrochemical environment can cause additional difficulties if the voltage applied between the probe tip and the electrode induces electrolysis, which can mask the tunnelling current. To minimize this problem the tip (except its very end) is coated in a non-conducting wax or polymer to reduce the area available for electrolysis.

An elegant example of electrochemical STM has recently been reported by Weaver (1992) where the oxidation of sulphide on a gold (111) surface is monitored as a function of potential at the atomic/molecular level. Figure 5.11 shows an STM image of the Au (111) surface in an aqueous solution containing Na_2S, where the electrode is potentiostatted at a voltage insufficient to induce sulphide electro-oxidation. The image recorded shows an adlayer of adsorbed sulphur on the gold surface.

When the electrode potential was altered to 0V (vs SCE), which is sufficient to induce electro-oxidation, a substantially altered image was observed (Fig. 5.12). A close inspection of the image reveals a series of structures on the surface each of which appear to have eight individual tunnelling current maxima. The authors interpreted each of these maxima as individual sulphur atoms on the surface, implying that the S_8 species had been formed at the electrode. This type of behaviour had been implicated from previous Raman experiments performed under similar conditions where the spectra recorded indicated the formation of S_8.

The authors were able to identify the sulphur adlayer structure as a hexagonal ($\sqrt{3}\times\sqrt{3}$) lattice.

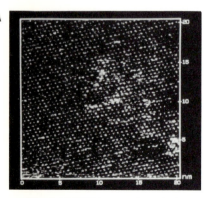

Fig. 5.11 STM image of a gold (111) surface in an aqueous solution of Na_2S.

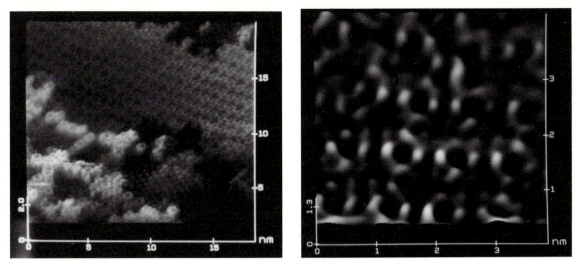

Fig. 5.12 STM image of a Au (111) surface (left) in an aqueous solution of Na_2S potentiostatted at 0 V (vs SCE) and (right) following removal from the electrolyte solution.

The use of tiny probe tips to interrogate solid/liquid interfaces is not confined to the STM experiment. For example, atomic force microscopy, uses a similar experimental arrangement as STM but measures the variation in the force (of the order of 10^{-9} N) between the probe and the substrate. A further variation is that of scanning electrochemical microscopy (SECM) which is considered next.

5.3   Scanning electrochemical microscopy

Scanning electrochemical microscopy (SECM) has recently been described (Bard 1989) as a new method for imaging surfaces in solution. A schematic diagram of the important components of the apparatus is shown in Fig. 5.13. Note that the substrate and probe form part of an electrochemical cell together with reference and auxilliary electrodes.

The imaging can be described by considering a simple electrode reaction which is induced to occur at the microelectrode

$$A(aq) \pm e^-(m) \rightleftharpoons B(aq)$$

The species A is present in the solution phase above the substrate and the potential of the probe is held at a value that induces the diffusion-limited conversion of species A to species B. The current flowing at the probe rapidly establishes a steady-state value, and, if the tip–substrate distance is sufficiently large, the resulting concentration profile around the probe is hemispherical (Fig. 5.14). However, as the probe is lowered towards the surface the concentration profile around the probe tip is perturbed by the substrate below and the observed current is altered from the unperturbed value.

The technique can be used to investigate both conducting and non-conducting solids. In the case of a non-conducting solid, when the probe

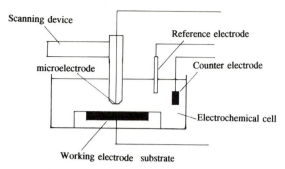

Fig. 5.13 Schematic of SECM apparatus.

approaches the surface, diffusion of the reactant to the probe is partially blocked by the sample and the current detected at the electrode decreases. This is termed hindered diffusion (Fig. 5.15). For the case where the sample is conducting the potential of the substrate is set so that any B that reaches the surface is electrochemically transformed to A. Consequently, more A is produced than for the corresponding situation in bulk solution and the current detected at the probe increases. This is termed feedback diffusion (Fig. 5.16).

At sample–probe distances up to several times the radius of the microelectrode probe the current detected at the probe is very sensitive to distance. By scanning the probe tip over the sample at a constant height and simultaneously recording the current a topographical image of the surface can be obtained.

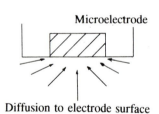

Fig. 5.14 Hemispherical diffusion.

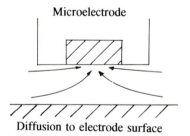

Fig. 5.15 Hindered diffusion.

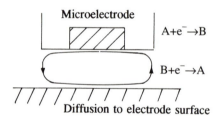

Fig. 5.16 Feedback diffusion.

As the technique may be used with a wide range of electroactive species (A/B) it is possible to chose materials that may interact with the surface. The technique can therefore also be used to investigate the chemical properties of the substrate surface. For example, the species A may be capable of reaction at the substrate surface, depleting the concentration of A near the probe. A drop in the measured current can then be interrogated to ascertain the nature of the reaction.

An illustration of the versatility of this device is shown in Fig. 5.17(a). In this mode the SECM is employed as a microstructure fabrication tool. A current is passed between the tip of an ultra microelectrode and a metal substrate. The electrodes are separated by a NAFION film (NAFION is a conductive polymer). The current that passes through the film is essentially controlled by

(a)

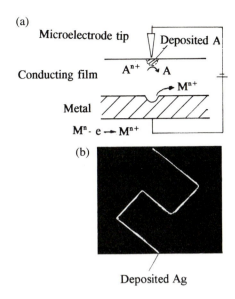

Fig. 5.17 (a) Schematic of the processes occurring in the deposition process at the nafion film/microelectrode interface. (b) The resulting image of metal deposited on the nafion film.

the area where the microelectrode touches the film. When an appropriate potential is applied to the ultra microelectrode a reduction reaction is induced at the electrode/film interface with the 'complementary' process occurring at the substrate. The result is electrochemical deposition at the nafionfilm/microelectrode interface and etching at the metal substrate. Careful positioning of the probe enabled the production of thin metal lines on the film (Fig. 5.17(b)).

5.4 Spectroscopic detection of homogeneous electrogenerated species

The previous sections have focused on the characterization of the electrode–solution interfacial region. The problem considered next is that of the molecular identification of solution phase intermediates and products resulting from electrode reactions. Again the merits of in-situ detection over experiments that require sample transfer between electrochemical cell and a spectrometer will be apparent.

5.5 Ultraviolet/visible spectroscopy

One of the first *in situ* combined electrochemical/ spectroscopic techniques to be investigated employed UV/visible detection. Solution-phase spectra of organic radicals generated at an electrode were recorded. Typically, organic intermediates/products possess additional absorption bands not observed in the parent molecule. These arise since new electronic transitions are possible once the parent molecule has undergone electron injection or removal. The new bands can be used to fingerprint the electrogenerated species. By way of example (Fig. 5.18) shows the absorption spectrum of anthraquinone and the corresponding radical anion generated at a gold electrode.

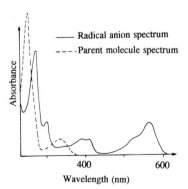

Fig. 5.18 The UV/vis spectrum of anthraquinone(– – – –) and its radical anion (———).

Early research (Kuwana 1964) used optically transparent electrodes such as antinomy-doped tin oxide deposited on a silica substrate. The electrode allows the analysing beam of a UV/visible spectrometer to pass directly through the electrode and the cell before detection in the usual manner, as shown in Fig. 5.19.

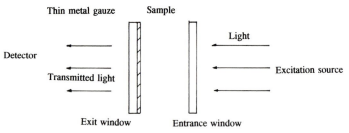

Fig. 5.19 Schematic representation of an optically transparent electrode for UV/visible spectroscopy.

This approach has been used to monitor the electrochemical oxidation of *o*-tolidine in aqueous hydrochloric acid. The formation of a dication was demonstrated

$$H_3\overset{+}{N}-\!\!\!\!\bigcirc\!\!-\!\!\bigcirc\!\!-\!\!\overset{+}{N}H_3 \;\rightleftharpoons\; H_2\overset{+}{N}=\!\!\!\!\bigcirc\!\!=\!\!\bigcirc\!\!=\!\!\overset{+}{N}H_2$$

$$+\ 2\ e^-\ +\ 2\ H^+$$

In addition to the gathering of spectra, useful information may be gleaned if the intensity of absorbance is monitored at a fixed wavelength as a function of time after the electrode is subjected to a potential step or pulse. For example, suppose the electrode potential is jumped from a potential where no current flows to one at which some electrogenerated product is formed. The current flowing is initially very large but gradually drops as parent material is consumed (see Section 3.6). The spectroscopic signal intensity recorded from the electrogenerated product gradually grows with time until a steady state is obtained (Fig. 5.20). The shape and size of the intensity/time profile is characteristic of the reactivity of the product and such measurements may be used to give estimates of the lifetime of electrode intermediates.

Fig. 5.20 A schematic plot of the growth of spectroscopic signal intensity as a function of time for an electrogenerated species.

5.6   Infrared spectroscopy

Compared to UV/visible spectroscopy the use of infrared (IR) spectroscopy presents additional difficulties. Specifically the overall sensitivity of IR is lower than UV/vis and the solvents used in electrochemical experiments (such as water) often possess molecular vibrations that give rise to strong absorption bands in the IR region of the electromagnetic spectrum. To overcome these limitations a sensitive technique was developed called electrode modulated infrared spectroscopy (EMIRS). Light reflected from an electrode contained in a thin layer cell is detected and this provides information about species close to, or adsorbed on, the electrode surface. The thin layer cell (Fig. 5.21) reduces the path length of the analysing beam in the solution and therefore lowers absorption signals due to the solvent.

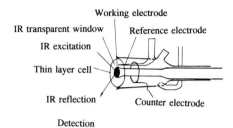

Fig. 5.21 Schematic of infrared electrochemical apparatus.

However, to gain adequate sensitivity towards electrode-related material, the electrode potential is pulsed between a potential at which no electrode reaction occurs and a value that induces electrochemical reaction. As a result the spectroscopic signal due to the species of interest fluctuates with a frequency identical to that of the potential pulse. In contrast, the signal due to the solvent or bulk solution species remains effectively constant. This provides a method for discriminating between electrogenerated and other species, and by constraining the detection to signals with the frequency of the potential pulse the resulting spectra reflect solely products of the electrode reaction. After analysis of the data it is possible to detect sub-monolayer quantities of material adsorbed at the electrode surface or in a region of solution near the electrode.

In a typical application (Berwick 1980) EMIRS has been employed to probe the adsorption of indole on a platinum electrode. Indole adsorbs on to platinum at around 0 V but can be desorbed at about 1.2V (versus a Ag/Ag$^+$ reference electrode). Pulsing the potential between these values enabled an absorption spectrum to be recorded of the adsorbed molecule. The latter differed from corresponding solution-phase spectrum only in the N—H stetching region. It was consequently inferred that the indole was bound to the electrode surface via an interaction with the nitrogen atom.

In a further application the EMIRS approach has been used to reveal the formation of a radical cation in the solution phase following the oxidation of thianthrene in acetonitrile solution (Fig. 5.22).

5.7 Raman spectroscopy

Electrochemical reactions have also been studied using Raman spectroscopy, which relies on the inelastic scattering of light due to interaction of an

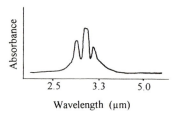

Wavelength (μm)

Fig. 5.22 The EMIRS spectrum of indole adsorbed on a platinum electrode.

incident excitation beam with vibrational energy levels in molecules. The excitation source is usually an ultraviolet/visible laser and consequently fewer problems due to solvent absorption occur using this technique. A typical experimental arrangement is depicted in Fig. 5.23.

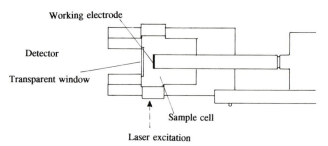

Fig. 5.23 Schematic of electrochemical apparatus employed to study Raman spectroscopy.

The technique has been used to study the oxidation of hydroquinone to the benzoquinone (Clarke 1974).

Transient electrochemical experiments were performed and the intensity of new bands from the quinone recorded. All the solution-phase species in the scheme above are Raman active so that variation of the applied potential enabled the electrochemical reaction sequence shown to be established.

5.8 Luminescence spectroscopy

The identification of electrochemically generated species has also been performed using luminescence techniques. In this technique a conventional spectrometer is used in conjunction with a flow cell so that electrogenerated intermediates are swept away from the electrode and the luminescence analysed downstream of the electrode (Fig. 5.24). The luminescence from the sample is detected at 90° to the excitation source. Using this approach Compton (1994) showed the dimerization reaction of the electrochemically generated PPD$^+$ cation radical has been identified

$$PPD(aq) - e^-(m) \rightleftharpoons PPD^+(aq)$$

$$2\,PPD^+ \rightleftharpoons (PPD^+)_2$$

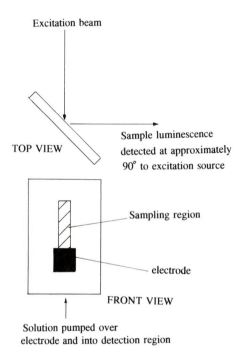

PPD

The technique itself is particularly well suited to the investigation of photo-electrochemical processes, since it can yield information about excited states of reactive species and their lifetimes.

Excitation beam

TOP VIEW

Sample luminescence detected at approximately 90° to excitation source

Sampling region

electrode

FRONT VIEW

Solution pumped over electrode and into detection region

Fig. 5.24 Schematic of electrochemical luminescence apparatus.

5.9 Electron spin resonance spectroscopy

The value of electron spin resonance (ESR) in the investigation of electrochemical processes arises from the fact that the technique is exclusively sensitive to molecules that possess unpaired electrons and are therefore paramagnetic. The electron has two spin states ($\pm 1/2$), which, when placed in a magnetic field become non-degenerate, and it is possible for the electron to absorb energy and jump from the lower to the higher energy spin state. This energy gap corresponds to microwave frequencies in the typical fields employed by X-band spectrometers.

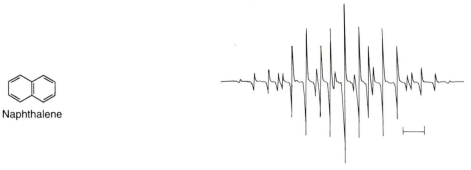

Fig. 5.25 The ESR spectrum of the naphthalene radical anion.

Fig. 5.26 The ESR spectrum of semifluorescein.

$$F(s) \quad + \quad e^-(m) \quad \rightleftharpoons \quad S^{\cdot}(s)$$

$$S^{\cdot}(s) \quad \overset{h\nu}{\longleftarrow} \quad S^{\cdot\cdot}(s)$$

$$S^{\cdot\cdot}(s) \quad + \quad H^+(s) \quad \rightleftharpoons \quad SH^{\cdot\cdot}(s)$$

$$SH^{\cdot\cdot}(s) \quad + \quad S^{\cdot}(s) \quad \overset{slow}{\rightleftharpoons} \quad F(s) \quad + \quad LH(s)$$

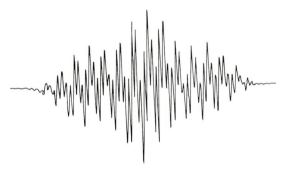

The products of electrode reactions involving organic molecules often include radical species and ESR measurements made simultaneously with electrolysis enable their selective and sensitive detection. For example, in Fig. 5.25 the radical anion is detectable using ESR spectroscopy but the parent naphthalene molecule is invisible.

Recent experimental strategies have utilized micro- and macroelectrode arrangements. In the case of macroelectrodes forced convection has been exploited to investigate complex mechanistic processes such as the photore-duction of the dye fluorescein. Combined voltammetric and spectrometric data allowed the reaction mechanism to be established (Compton 1988) and the detection of the semi-fluorescein molecule S.

Bibliography

Bewick A., Kunimatsu K. and Pons B.S. (1980), *Electrochim. Acta.*, 25, 465.

Binnig G., Rohrer M., Gerber C. and Weibel E. (1983), *Phys. Rev. Lett.*, 50, 120.

Clark J.S., Kuhn A.T. and Orville-Thomas W.S. (1974), *J. Electroanal. Chem.*, 54, 253.

Compton R.G., Coles B.A. and Pilkington M.B.G., (1988), *J. Chem. Soc., Faraday Trans.* 1, 84, 4347.

Compton R.G. and Wellington R.G. (1994), *J. Phys. Chem.*, 98, 270.

Denault G., Lee C., Mandler D., Wipf D.O. and Bard A.J. (1990), *Acc. Chem. Res.*, 23, 357.

Gale R.J. (1988), *Spectroelectrochemistry, Theory and Practice*, Plenum, New York.

Gao X.P., Zhang Y. and Weaver M.J. (1992), *J. Phys. Chem.*, 96, 4158.

Hubbard A.T. (1989), *Comprehensive Chemical Kinetics,* 28, 1.

Kuwana T., Darlington K.R. and Leedy D.W. (1964), *Anal. Chem.*, 36, 2023.

Lu F., Salaita G.N., Laguren-Davidson L., Stern D.A., Wellner E., Frank D.G., Batina N., Zapien D.C., Walton N. and Hubbard A.T. (1988), *Langmuir*, 4, 637.π

Index

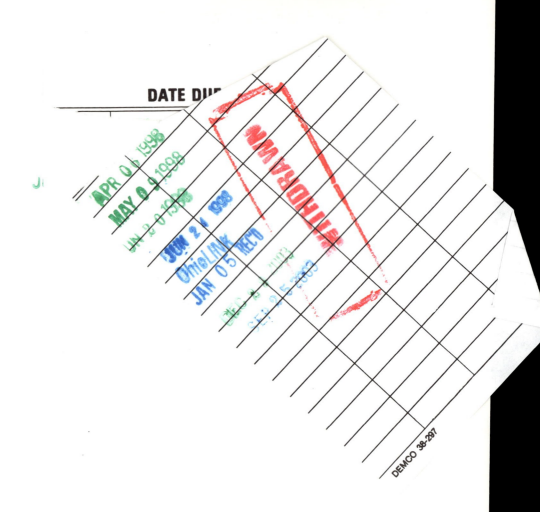